BEGINNER'S GUIDE TO ELECTRICITY & ELECTRICAL PHENOMENA

BY W. EDMUND HOOD

TAB BOOKS Inc.
BLUE RIDGE SUMMIT, PA. 17214

FIRST EDITION

FIRST PRINTING

Copyright © 1983 by TAB BOOKS Inc.

Printed in the United States of America

Reproduction or publication of the content in any manner, without express permission of the publisher, is prohibited. No liability is assumed with respect to the use of the information herein.

Library of Congress Cataloging in Publication Data

Hood, W. Edmund
    Beginner's guide to electricity and electrical
phenomena.

    Includes index.
    1. Electric engineering. 2. Electronics. I. Title.
TK146.H586   1984   621.3   83-4945
ISBN 0-8306-0507-X
ISBN 0-8306-1507-5 (pbk.)

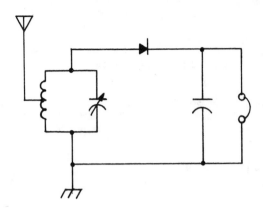

Contents

Introduction   v

1 Static Electricity and the Early Years   1
Atoms and Electrons—Conductors and Insulators—Electrical
Charges—Static Electricity—A Simple Electroscope—The
Electrophorus—The Gold-Leaf Electroscope—The Awful Jar at
Leyden—Industrial Uses of Static Electricity—Static in Nature

2 Direct Current and Simple Circuits   26
The Voltaic Cell—The Storage Cell—Voltage, Current, and
Resistance—The Simple Circuit—Series and Parallel Circuits

3 Magnets   49
Basic Magnetic Principles—Induced Magnetism—How a Com-
pass Works—Tracing the Magnetic Field

4 Electromagnetism   59
A Galvanometer—Right-Hand Rule—Permeability—Simple
Solenoid—The Telegraph

5 The Mysterious Alternating Current   76
Electromagnetic Induction—Motors—A Simple Generator—Sine
Waves—Transformers—Electric Sparks

6 The Basics of Home Wiring   93
Safety Grounds—Power Cords—Basic Circuits—Wiring
Types—Boxes and Fixtures—Fuses and Loads—Service

7 Electricity and Sound   **110**
The Story of the Telephone—Telephone Reproducer—A Carbon Microphone—Sound-Powered Telephone—Phonographs

8 The Electronic World   **125**
Early Diode Experiments—Vacuum Diode—Solid-State Diodes—Circuits and Symbols—The Half-Wave Rectifier Circuit—A Bridge Circuit—Filtering the Dc

9 Vacuum Tubes and Transistors   **138**
The Vacuum Tube—Transistors—Transistor Types—Amplifier Techniques—Vacuum-Tube Amplifier Experiment—Transistor Audio Amplifier

10 Radio Communication   **155**
Early Systems—Voices—Resonance—Receivers—Transmitters—Television

11 Digital Logic   **178**
Transistor Switching—Gates—An Event Monitor—Binary Numbers—Exclusive-OR Gate

12 Computers   **205**
The CPU—Memory—The ALU—Programming Basics—The BASIC Language

Appendix Where to Get the Goodies   **225**

Glossary   **227**

Index   **248**

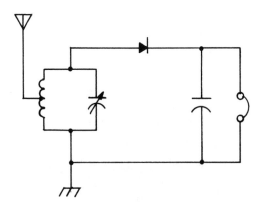

Introduction

From its beginning sometime during what we now call the gaslight era, the range of methods we use to put electric power to work has not stopped growing. Moreover, it's not going to stop for a long, long time. The battery, the electromagnet, and the generator have been almost as important in the progress of mankind as the invention of the wheel.

The scope of knowledge is today as widespread and as specialized as the field of medicine. A TV expert might know nothing about computers; a ship's radio operator may not know the first thing about fixing electric motors, and none of these need necessarily know the finer points about power distribution in a building. On the other hand, each can know a little bit about the other's field.

It is a nearly impossible task to condense the entire scope of electrical knowledge that any one of these men might have after thirty-odd years' experience into one single book. To even skim lightly over so wide a scope is a lengthy task (to which a very patient publisher will attest).

This book, then, is not intended to give complete coverage to the entire field of electric phenomena. It is, rather, the bottom at which a beginner may start, and a handy reference to which the old-timer can turn if he forgets something.

This book is dedicated in affectionate memory to
Arthur Fred Hood
1904-1981

He witnessed much of the history herein outlined. He was one
of the curious. He investigated, he tinkered, he experimented. He
reminisced, he taught, he encouraged.
And so I am what I am.

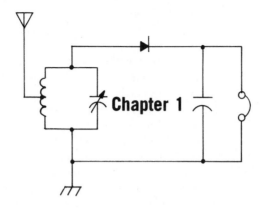

Static Electricity and the Early Years

It is just a hundred years since the invention of the incandescent lamp, thus, as a society, we have been "electrified" for less than one century. Although we still don't know all there is about this wonderful power, we understand its behavior well enough to perform daily miracles with it.

The existence of electricity has been known for over 2600 years. It was in the seventh century B.C. that a Greek named Thales of Miletus noted some peculiar properties of amber. This rosinous material, which the Greeks called *elektron*, would attract small bits of material after being briskly rubbed. The particles would cling for a moment or two, and then fly away. For a great many years, that was all that was known about electricity.

We had to get out of the dark ages before we could learn much about electricity. Before that, research was badly hampered by superstition. In the thirteenth century, for example, Roger Bacon began making a detailed study of materials that displayed electrical properties. He was labeled a blasphemer and put into prison.

Three hundred years after Bacon finally got out of prison, William Gilbert began to list some of the materials, and named the force *vis electrica*. His work was widely published, and even drew praise from the great Galileo. There then began a general fad of electrical fun. Machines were invented that would produce fairly heavy electrical charges, and such stunts as suspending a small boy from the ceiling to watch him attract splinters, or charging up a pretty girl and having her boyfriend kiss her, were commonplace.

All of this horseplay, whether done for entertainment, or out of genuine scientific interest, contributed bits and pieces to general knowledge. Still, it was to be several centuries before the real secrets could be unlocked. Scientists had to grope their way along. There were no precedents to which they could refer, nor were there any instruments to measure the charges they were working with, unless the researchers themselves invented them.

In spite of all the impediments, knowledge of electricity began to accumulate. Around 1729, Stephen Gray discovered that the electric "fluid," as it was called, could flow through some materials. Shortly thereafter, a Frenchman, Charles Dufay, noted that two kinds of electricity existed which could neutralize one another. Benjamin Franklin named them *positive* and *negative*.

Dufay and his partner, Nollet, conducted numerous spectacular experiments to determine how fast electricity flowed. Enlisting the aid of a monastery, he lined up the monks in a column over three kilometers long and gave them an electric jolt to see if they would all jump at the same time. They did.

In coining the terms "positive" and "negative," Franklin made one wrong guess. Positive, you see, implies a surplus of something, while negative implies a shortage. Of the two types of charges, we now know that the opposite is true. We can easily forgive Franklin his error, for we now know something that he couldn't possibly have known.

ATOMS AND ELECTRONS

The whole secret lies deep within the tiniest of particles, the atom (Fig. 1-1). Whirling about a nucleus are minute particles of energy called *electrons*. Each electron has a negative electrical charge. When some of the electrons are removed from the atoms of a material, that material displays what we call a positive charge; when additional electrons are forced into the material, it displays a negative charge (Fig. 1-2).

CONDUCTORS AND INSULATORS

Most materials are in one or the other of two classes—they are either conductors or insulators. A *conductor* is a material through which electrons can easily flow from atom to atom. An *insulator* is a material through which electrons cannot flow. Typical conductors

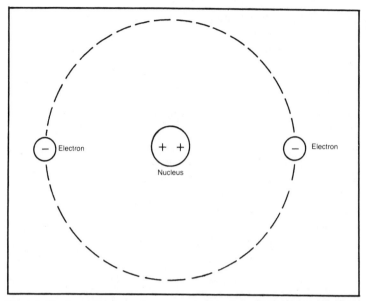

Fig. 1-1. The inner structure of an atom consists of a nucleus with one or more electrons orbiting around it. The electrical charges of the nucleus balance those of the orbiting electrons.

are most metals, carbon, and certain liquids. Insulators include glass, rubber, wood, plastic, and almost all nonmetallic solids.

Since electrons cannot flow through an insulator, they accumulate on its surface, thereby forming a static charge. Electrons

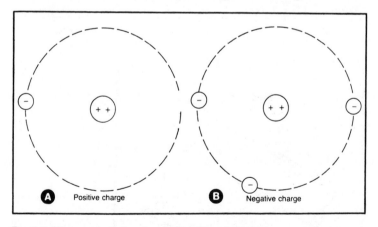

Fig. 1-2. When one electron is missing, the substance has a positive charge (left); when an extra electron is present, a substance has a negative charge (right).

can be forced onto a material, or taken away, by rubbing it with something that produces friction (Fig. 1-3). Some materials give up electrons when rubbed, producing a positive charge; other materials pick up electrons, producing a negative charge.

Many materials will conduct very slightly when they are wet. (Witness the fellow who turns on the light while standing in the bath tub!) Therefore, on damp days when the air is very humid, static charges do not accumulate nearly as well as they do in a very dry environment. You've probably noticed that you often get unexplained electrical jolts off the car door or from the water pipe during the winter. These are from static electricity. You don't feel them in the summer because the summer air is more humid, and the charges quickly bleed away.

Because charges "bleed" away in humid summer air, the experiments that follow may not work too well during the summer. In fact, in a tropical or semitropical climate, they may not work at all.

ELECTRICAL CHARGES

Let's try an experiment that will produce some electrical charges. You will need the following equipment:

Glass rods (2)
Sealing wax sticks (2)
Silk handkerchief

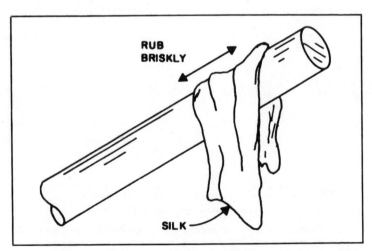

RUB
BRISKLY

SILK

Fig. 1-3. Electrical charges can be produced in a variety of substances by friction.

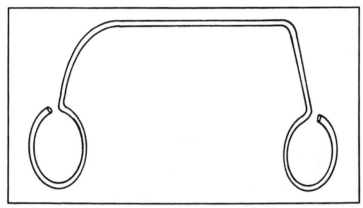

Fig. 1-4. A wire bracket or stirrup to hold the glass rod can be fashioned from a paperclip.

Flannel (scrap)
Paper clips
String
Sawdust (a small quantity)

Procedure. Rub a glass rod with a silk handkerchief. Rub briskly as if you were polishing it. Hold the glass rod close to the sawdust. Note what happens.

Bits of sawdust will fly up to the glass rod. They will cling for a moment, then fly away.

Rub a stick of sealing wax with the flannel. Hold it close to the sawdust. Note the results are the same as with the glass rod. (Note: If you are doing this in a warm room, you may want to put the sealing wax into a freezer for a while beforehand. This will ensure that the sealing wax is hard enough to withstand the rubbing without softening.)

Electrify the glass rod a second time. Hold it in a stream of running water for a moment, and again bring it close to the sawdust. Note what happens.

Nothing happens.

Bend a paperclip into a stirrup as shown in Fig. 1-4. Be sure the bends will fit around the glass rod without having to force it. Hang the stirrup on a string so that the glass rod will swing freely.

Now electrify the glass rod as you did before, and place it into the stirrup. Electrify the second glass rod and bring it close to the first (Fig. 1-5). Do the same thing using sealing wax instead of glass. Note what happens.

5

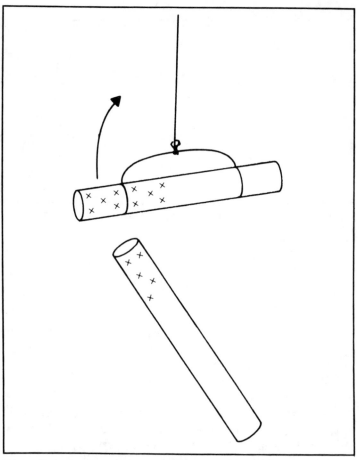

Fig. 1-5. Bring a positively charged rod close to another having a positive charge. They will push each other away.

In either case, the hanging material will swing *away* from the second piece.

Now electrify and hang up a glass rod again. Electrify a piece of sealing wax and bring it close to the glass rod. Note what happens.

The glass rod will swing *toward* the sealing wax.

Explanation. As we mentioned at the start of this chapter, the secret of this mystery lies within the very atoms of the substance; particularly with those atomic particles called electrons. The molecules of some substances are so structured that some of the electrons in their atoms are unstable, and therefore easily lost. The atomic structure of some other substances can easily have extra electrons forced into them. Substances that easily lose electrons will be referred to as *donors*, while substances that easily accept electrons will be referred to as *acceptors*.

When you rub a donor material with an acceptor material, as you did when you rubbed the glass with the silk, the donor loses many electrons to the acceptor. Thus the glass lost electrons and, since electrons are negative particles, became less negative than it should have been. This condition is called a *positive* charge.

In the same manner, the sealing wax picked up a lot of electrons from the flannel and gained a *negative* charge, representing a surplus of electrons.

A surplus of electrons is called a *negative* charge; a shortage of electrons is called a *positive* charge. Now comes the part where nobody has ever discovered why. Electrons will not tolerate even being close to one another. Moreover, they are strongly drawn to places where electrons belong but are missing. Nobody knows exactly why. That's simply what they do.

Electrons flow very easily through some substances. Through others, they either flow with great difficulty or not at all. The glass rod had electrons missing. The electrons on the bits of sawdust were strongly drawn toward the void in the glass (Fig. 1-6A). Because electrons couldn't flow through the air, so they took the sawdust bits with them right up to the glass rod. Once they were there, some of the electrons in the sawdust chips were drawn into the glass rod. The chips were so small that only a small amount of electrons were given up in comparison with the amount that had been earlier lost by the glass. Therefore, although the charge in the glass was lessened very slightly, a considerable positive charge was

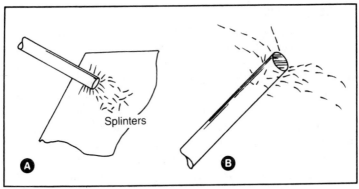

Fig. 1-6. A charged rod will pick up splinters or other small particles (A); the charge transfers to the splinters, then they fly away (B).

built up in the sawdust, until glass and wood repelled one another (Fig. 1-6B).

With the sealing wax, you produced a surplus of electrons, which we call a negative charge.

When you charged up a glass rod and then held it in running water, the water conducted electrons from the ground to the glass until its charge was neutralized. Then the glass rod no longer attracted the wood chips.

When you brought the two positively charged glass rods close together, they repelled one another. The same thing happened when you brought the two negatively charged pieces of sealing wax together. However, when you brought the positive glass close to the negative sealing wax, they attracted one another. That illustrated the most important of electrical rules: *Like charges repel; unlike charges attract.*

It may seem strange to you that the word "negative" is used to indicate a surplus of electrons, when in other applications the same word indicates a shortage. The word "positive" seems to be similarly misapplied. We owe this particular usage to early electrical experimenters who, around the time of the American Revolutionary War, observed that there appeared to be two kinds of electricity that appeared to be opposite to one another, and were capable of canceling each other. It was presumed that one represented a surplus of that unknown stuff (they called it electric fluid, giving rise to the expression "juice") while the other represented a lack of it. Benjamin Franklin applied the terms "positive" and "negative" to define the two. It wasn't his fault that he guessed wrong as to which of the two was the surplus and which was the shortage.

STATIC ELECTRICITY

Of the various phases of electrical science, static electricity is the most common in nature, and the least useful commercially. You can, however, have a lot of fun with it, using a miminum of equipment, and it can delight the small fry. For example, a toy balloon, rubbed on the top of your head, will cling to the wall or window. A piece of paper, laid on the table and rubbed with a plastic bread wrapper (wadded up), will develop a strong charge and stick to almost any smooth surface. These can keep any youngster wide-eyed when done with a flourish or two and a few "abracadabras." For sheer fantasy, however, nothing can really beat the electric dance.

The Electric Dance

Here is a simple trick that will demonstrate electrostatic attraction. It can be set up and operated safely by any youngster without supervision.

You will need:

Frame (about 10 inches square and 2 inches deep). This can be homemade, or you can use four wood blocks in its place.

Window glass (8 by 10 inches or whatever size is handy). Plexiglas will also work.

Wood splinters, paper scraps, or tiny paper dolls (1 handful).

Flannel or silk (9 by 12 inches).

Support the glass 1 or 2 inches over a tabletop, with the wood chips or paper bits under it. Rub the glass briskly with the flannel and watch the chips dance (Fig. 1-7). They will jump to the glass and drop away several times. Some will cling to the glass longer than others; some will move about on the under surface of the glass in a wild-looking dance.

This experiment may not work well on a damp or muggy day. You may have better results in the summertime if the room is air conditioned. The best results will be obtained in the winter, especially when the weather outside is cold and clear. Results in the summertime can be improved somewhat by first putting the flannel in a clothes dryer for a few minutes.

A SIMPLE ELECTROSCOPE

An electroscope is a device that detects electric charges. The earliest and simplest models consisted of a pith ball hanging from a support on a piece of thread. Pith is a soft material contained inside the stems of certain plants, such as elder. When dried it becomes

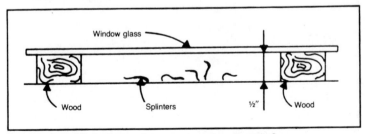

Fig. 1-7. The electric dance. Can you guess how it works?

very light in weight. If it seems to be too much trouble getting pith, you can use a piece of the breakfast cereal called Kix. I have tried that, and it does work. You will need:

Wood block (a scrap of two by four, four inches long)
Wire coat hanger
Thread, about 8 inches
Pith, cork, or Kix
Aluminum spray paint

Drill a hole, large enough to receive the coat-hanger wire, into the wood block. Bend the wire into an L shape, and insert it into the block, as shown in Fig. 1-8. Using a sewing machine needle, pass the thread through the pith ball, and tie a knot in that end. Tie the other end to the end of the wire. Spray the ball lightly with the aluminum paint. Be careful not to spray paint the thread as well.

To use the electroscope, simply bring the object suspected of being charged close to the ball. The ball will swing toward the object if it is charged. If the ball touches a charged object, it will cling for a moment and then fly away. Try it, using a charged glass rod. Once the pith ball touches the rod, the ball will have the same charge as the rod, and the two will repel one another. If you then grasp the ball

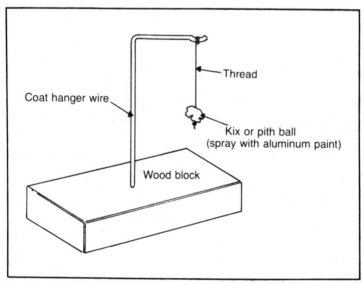

Fig. 1-8. A simple electroscope. This is *not* a breakfast-cereal advertisement. Kix just happened to be the substance of that kind I remembered by name!

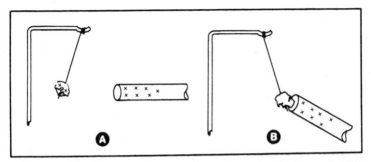

Fig. 1-9. As with the splinters in the earlier experiment, the electroscope will be attracted to the rod (A). It will stick until it picks up some of the charge, then it will be repelled (B).

lightly between your fingers, it will lose its charge and again will be attracted by the rod.

Charge up the ball with the glass rod, and then bring a charged-up plastic handle close to it. The ball will be attracted to the plastic until it touches, and then assume the charge of the plastic after which the two will repel. Now, having an opposite charge as the glass rod, the ball will swing to the rod, touch it, and change charges again (see Fig. 1-9). As with the other experiments, you will probably get the best results when the air is very dry.

THE ELECTROPHORUS

This was once a very important laboratory device, but today it is an all-but-forgotten curiosity. In the mid 1700s, when there was a lot of experimenting going on, a source was needed for electric charges. The electrophorus, invented by Volta, was the best answer. Here are the materials you'll need to make one:

Small aluminum pie tin
Rosin or sealing wax (a possible source of rosin is a music store)
Metal disc, small enough to just fit into the pie tin
Plastic handle

Melt the rosin and pour it into the pie tin until it's about half an inch or more deep. Young experimenters may need adult supervision here. Set the dish of rosin aside where it can harden undisturbed.

Fasten the handle to the center of the metal disc, using epoxy rosin, "Krazy Glue," or similar (see Fig. 1-10).

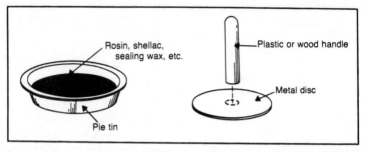

Rosin, shellac, sealing wax, etc.

Plastic or wood handle

Metal disc

Pie tin

Fig. 1-10. Simple as it was, the electrophorus was once an important laboratory tool.

To use the electrophorus, polish the surface of the rosin with a piece of warm flannel. Then place the disc on the rosin, momentarily put your finger on the disc, and then lift the disc, Fig. 1-11. It will be found to contain a heavy static charge.

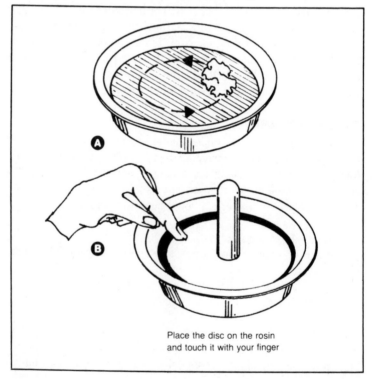

A

B

Place the disc on the rosin and touch it with your finger

Fig. 1-11. The rosin in the dish is first charged by wiping it briskly in circular motions (A). Then you place the disc on the rosin and touch it with your fingers (B). Presto! It's charged!

The charge on the electrophorus disc is strong enough to produce a spark if you bring your finger close to it. The disc can be charged up over and over with only an occasional recharging of the rosin.

Electric Bugs

Charge up the electrophorus in the usual manner. Place a number of pith balls or pieces of Kix on the disc. (It will work better if the pith balls or Kix are sprayed with aluminum paint.) As you lift the disc from the electrophorus, the small pieces will leap away like so many frightened bugs (Fig. 1-12).

THE GOLD-LEAF ELECTROSCOPE

Simple as this device may be, it still is perhaps the most sensitive device in existence for detecting minute electrical charges without going to great expense. It consists of two small strips of gold leaf (if you can get it) or aluminum foil hammered very thin, mounted in a jar, with a large conductor leading to the outside. When any charged object touches the outside end of the conductor, the strips of foil repel one another and spread apart.

To make a gold-leaf electroscope, you need the following materials:

A glass jar (of whatever size is convenient) with a metal cover

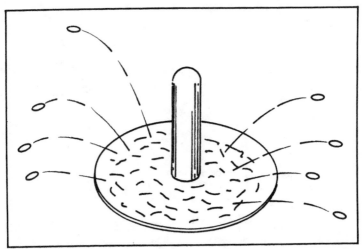

Fig. 1-12. Sprinkle small bits of sawdust on the disc. Watch them jump off.

Two inches of heavy copper wire

A 2½-inch strip of gold leaf, ¼-inch wide, or very thin
    aluminum foil

Remove the paper liner from the inside of the jar cap, exposing
bare metal. Remove the paint from the outside of the cap. Bend the
wire as shown in Fig. 1-13 so that it will hang down an inch or so
from the cap.

Gold leaf may be nearly impossible to get, unless you live
within reach of a supplier of jeweler's raw materials. Aluminum foil
can be used if you hammer it very thin. Place the foil on a flat metal
block, and pound it with a heavy hammer. Ideally, it should droop
like a limp dishrag.

Bend the foil double and attach it to the wire by crimping the
end as shown in Fig. 1-14. Solder the wire to the center of the cap
and screw the cap onto the jar.

To test the electroscope, charge a glass rod and bring it into
contact with the cap. The foil strip should spread apart, indicating a
charge.

To discharge the electroscope, simply touch the cap with your
finger. The ends of the foil strip should immediately come together,
indicating no charge.

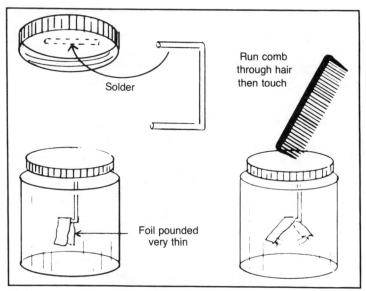

Fig. 1-13. This kind of electroscope is a bit tricky to make, but once you get it
working, it's very sensitive.

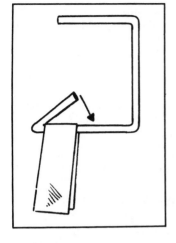

Fig. 1-14. You must pound the foil very thin. Then crimp it onto the end of the wire.

THE AWFUL JAR AT LEYDEN

Leyden is a small town about fifteen miles north of Rotterdam in Holland. During the mid eighteenth century, when experimenters all over the world were looking into electric phenomena, a scientist by the name of Pieter van Musschenbroek was a Leyden, conducting some investigations with his student, Cuneus.

At that time there was a theory that electricity was some kind of fluid. Musschenbroek was seeking a way to collect that fluid. Using a static electricity generator invented earlier by the English scientist Hawksbee, he connected an iron chain from the generator so that it dangled into a jar held by the student. After running the generator for a while, the jar acquired an electric charge, but there was no sign of any fluid. Musschenbroek knew that electric charges flow more easily through wet substances, so he put some water into the jar. Once again they set up the experiment and started the generator. Once again they watched as the generator ran. Still no sign of any fluid accumulating in the jar. The experiment was declared a failure. Cuneus reached up to disconnect the chain.

To Musschenbroek's amazement, Cuneus suddenly flew backward, equipment flying in all directions. The scientist rushed to the aid of his student. The young man was lying on the floor, pale as a ghost, and scarcely breathing. He had just received an electric shock such as no man had ever been known to have received before. Even though nothing had been observed accumulating in the jar, they had apparently stored quite a bit of the electric "fluid."

They had guessed far more accurately than one might expect. They had, in fact, stored a considerable amount of electric charge in

15

the jar. Electrons from the generator had flowed through the chain into the water. They gathered up against the glass, and the repelling force had pushed electrons out of Cuneus's hand, through his body, to the ground. When Cuneus grabbed the chain, the charges neutralized themselves by flowing through his body. The result was that he received quite a jolt. See Fig. 1-15.

When Musschenbroek wrote of his experiment, he said that it was extremely dangerous, and warned the scientific world not to try it. He didn't understand human nature very well; his warning was the perfect invitation for all the world to check out his discovery. Consequently, the Leyden jar was duplicated all over the civilized world.

The Leyden jar quickly became both a popular fad and an important laboratory tool. It is still in use today. There are dozens of them, in varying sizes, in the everyday transistor radio; there are

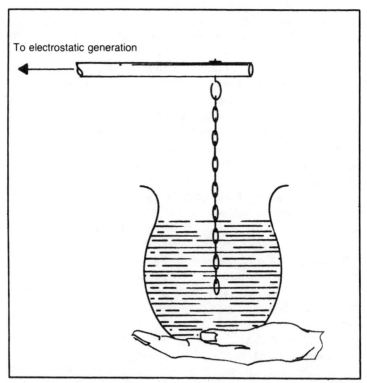

To electrostatic generation

Fig. 1-15. Poor Cuneus! He was probably the first man in history to get belted by a charged capacitor. He's now a member of a very large club. This is the contraption he used, called a Leyden jar.

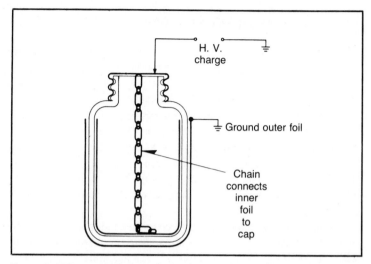

Fig. 1-16. All the warnings about the Leyden experiment only led to the improvement of the device. You can make this one.

hundreds in a TV set. They are now called capacitors, not Leyden jars. In the beginning, however, they were quite a novelty. In the Jules Verne novel, *20,000 Leagues Under the Sea,* Verne mentions Captain Nemo's use of devices similar to Leyden jars being used to kill sharks under water.

A Home-Brew Leyden Jar

The Leyden jar is so basic a thing that several versions can very easily be made without departing too far from the original design. First, let's try the jar-and-foil model. Here's what you'll need:

> One peanut-butter jar of convenient size (minus the peanut butter). Either a plastic or a metal cap is okay.
> Aluminum foil wrap.
> Metal chain long enough to reach from the cap to the bottom of the jar with an inch or so left over. Chain with ¼-inch links will be most workable.
> Stove bolt, small enough in diameter to pass through a link of the chain, ⅜ to ½-inch long, with nut.

Line the jar on the inside with foil, up to an inch or so from the cap. Do the same with the outside of the jar (see Fig. 1-16). Drill a hole in the center of the cap, of the proper size to clear the bolt, and bolt the chain to the cap.

Put the cap on the jar. If you've done it right, the chain should hang down and some of it lies on the foil at the bottom of the jar.

To really charge up the Leyden jar, you'll need the electrophorous described earlier in this chapter.

Temporarily connect the outside foil to a good ground, such as a water pipe. Charge up the electrophorous, and touch the disc to the bolt in the cap of the Leyden jar. You may do this several times, if you're brave.

Now the fun part. Hold the jar in one hand and touch the cap with the other. Depending on how heavily you've charged it, you can get quite a jolt.

It has been suggested that a much more efficient Leyden jar can be made by using a plastic jar and painting the inside and outside with conductive paint. Such paint is often available at electronics stores.

Hang the chain from the cap as in the previous model. I haven't actually made one this way, and my only question would be whether the plastic can withstand as high a voltage as the glass. Otherwise, there is no reason why it wouldn't work.

A Homemade Capacitor

One experiment we have tried, however, is the homemade capacitor. To really try it out, you will need a high-voltage power supply in the neighborhood of 100 volts. Such a thing can be borrowed from a local trade school, if they're in a good mood, or if you can win the sympathy of a local ham-radio operator, he or she might be able to help you. Otherwise you might want to wait until you get to Chapter 8 and build one (experienced adult supervision strongly recommended).

You also need a package of aluminum foil and a length of plastic sheeting or waxed paper.

Roll out the entire roll of foil onto the floor and cut it into two equal lengths.

Cover one length of foil with plastic sheeting, slightly wider and several inches longer than the foil. Lay the other strip of foil on top, see Fig. 1-17. Connect a piece of wire to each strip as shown, and then roll the whole thing up into a compact package, see Fig. 1-18.

Connect the two wires from the capacitor to the power supply and gradually increase the voltage. Without shutting off the supply,

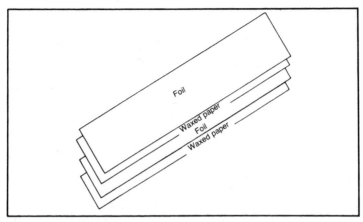

Fig. 1-17. Here's a simple, homemade capacitor that should hold a good charge—if it's made right.

remove the capacitor, *very carefully,* from the supply. It is now charged to the voltage of the supply. Be careful. Using one hand only, bring one of the wires into contact with the other. If you've done it right, you should get a strong spark at the moment of contact. If you make a mistake, you can get a strong shock.

To summarize this chapter, static electricity implies electrical charges that remain in one place until the moment of discharge, then do their thing and are gone. Except for the last experiment, current magnitudes in these experiments have been very small, and we have dealt primarily with voltage.

By voltage, we mean the difference of electrical charges between one object and another. The object having the lesser quantity of electrons is said to be the positive body, and the other negative.

It is perfectly possible for a charged object to be positive with respect to one body, and negative with respect to another at the same time.

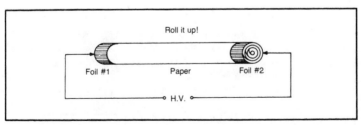

Fig. 1-18. The two foils should not touch one another. One foil sticks out one end, the other foil sticks out the other end.

1 Surface of a plate coated with a photo-conductive metal is electrically charged as it passes under wires.

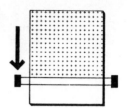

2 Plus marks represent positively charged plate.

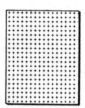

3 Original document is projected through a lens. Plus marks represent latent image retaining positive charge. Charge is drained in areas exposed to light.

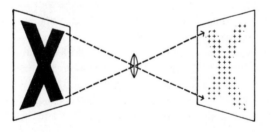

4 Negatively charged powder (toner or "dry ink") is applied to the latent image, which now becomes visible.

Fig. 1-19. How the Xerox electrostatic copying process works. (Courtesy Xerox Corp.)

5 Paper (or other material) is placed over plate and given positive charge.

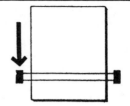

6 Positively charged paper attracts dry ink from plate, forming direct positive image.

7 Image is fused into the surface of the paper or other material by heat for permanency.

Xerography is basically an electrostatic process with a wide variety of applications. Commercial uses have been pioneered and developed by Xerox Corporation which now offers a broad array of xerographic copying and duplicating equipment as well as equipment used in diagnostic medicine.

Xerox also produces and markets computers and facsimile equipment, and has made a number of technological contributions related to the exploration of outer space. The company also is one of the world's largest publishers of educational material, serving both formal and informal markets and the Xerox label may be found on many library products and services such as microfilm and microfiche collections of literature and publications of all kinds.

Fig. 1-20. The first electrostatic copier, which came out in 1959 was a large, slow monster.

We have determined that some substances, generally metals, allow charges to move through them, while others resist the movement of charges. Charges flow easily through *conductors* and flow only with great difficulty through *insulators*. Most electric wires consist of a conductor covered with an insulator.

INDUSTRIAL USES OF STATIC ELECTRICITY

By its very nature, the use of static electricity in industry is limited. Some air-cleaning devices use static electric charges to remove dust or smoke from the air. Laboratory devices can store information in the form of a charge in a capacitor. However, the most widespread use of static electricity is in copying machines, and the following illustrations (Figs. 1-19, 20, and 21), generously furnished by the Xerox Corporation, shows how the process works.

Certain metals are known to be photo-conductive. That is, when they are exposed to light, they conduct electricity; otherwise they do not. A plate made of such a metal is given a static charge. The image of the document is then focused onto it. The bright portions of the image makes the plate conduct, leaving only the dark portions charged. The plate is then brought into contact with fine black powder, with a charge opposite to the plate. The powder is attracted to and held to those portions of the plate that are charged.

Next, a piece of paper is given a heavy charge of the same polarity as the plate. When the paper comes into contact with the plate, it picks up the black particles from the plate, and they stick to

Fig. 1-21. Xerox now has newer models that work much faster. All copiers of this sort depend on electrostatic attraction to work (Photos courtesy Xerox Corp.).

Fig. 1-22. The sequence of a lightning stroke. In A, a charge has built up between the clouds and ground. In B, probing "leader" strokes attempt to bleed off the built-up charge. When a pair of "leaders" meet, a sudden, massive flow of current results (C).

the paper in an exact replica of the original image. The paper is then heated to fuse the black particles permanently into place.

STATIC IN NATURE

Nature makes spectacular use of static electricity in the well-known phenomenom of lightning. The charges are generated by the movement of various particles of dust and moisture through the air as air masses of different temperatures come together. Vertical air currents within the clouds carry these charges upward where they accumulate until tremendous voltages are generated with respect to ground or to other parts of the cloud.

At a certain voltage, the resistance of air begins to deteriorate a little at a time. When the charges in a cloud exceed the voltage, the air begins to become conductive. Streams of conductive air descend from the cloud and ascend from the ground. If two of these streamers meet, current flows, instantly draining the accumulated charge. That current flow is what we know as a lightning stroke, see Fig. 1-22.

The voltage is much too high to be measured, but some experts have estimated it to exceed 100,000 volts for each foot of the lightning stroke. When we consider that the stroke can be as much as ten miles long, the voltage level becomes astronomical.

The magnitude of current has been estimated as high as a quarter million amperes. Even though the stroke lasts only about a quarter of a millisecond, the power dissipated can easily be calculated (using formulas given in a later chapter) to exceed that which the average household uses in a year. A single summer thunderstorm can dissipate more power than several nuclear bombs.

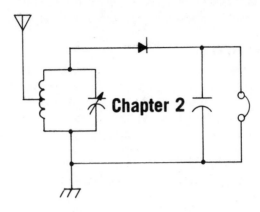

Direct Current and Simple Circuits

For many years electricity remained a laboratory curiosity with no real popular usefulness. The reason was quite simple; scientists could produce and discharge static charges, but these caused only instantaneous surges of current. Before this strange power could be utilized for the benefit of mankind, some means had to be discovered that would produce a sustained current flow—that is, a continuous source of power.

The breakthrough came gradually and began shortly after the American Revolution. An Italian biologist, Luigi Galvani, was dissecting a frog, and it happened that an electric machine, a static generator, was in use nearby. When the scalpel touched the freshly severed legs, they jerked convulsively as if they were alive. For that day and age, with knowledge as limited as it was, the sight must have been frightening. Some say that it was Mrs. Galvani who suggested that the electric machine had something to do with it. Whoever suggested it, Mr. Galvani began a research effort that was to last for years.

In the course of his experiments, Galvani discovered that the frog nerves were extremely sensitive detectors of electrical charges. Then, in September of 1786, another accidental step was taken along the road of discovery. Galvani had bound some frog legs with copper and, as a convenience, hung them on an iron railing nearby. When the exposed nerves accidently touched the iron, the legs jumped. There was no electric machine in the vicinity; there was no evidence of atmospheric electricity. (This was after Franklin's famous kite experiment.) Galvani found himself on the threshold of a whole new field of experimentation.

Before we judge Mr. Galvani too harshly, we must remember that he was a *biologist*, not a physicist. His primary field of study was that of living things. When he observed the reaction of the frog legs to the contact with the iron and the copper, he assumed that the electricity was coming from the animal tissue. Even though he quickly learned that the reaction occurred only when two different metals were used, he still thought he was on the way to discovering the secret of life itself—that the electricity being detected was a life-force coming from the animal tissue. He consequently directed his studies toward that end, and when he published his work, drew that conclusion.

THE VOLTAIC CELL

Among the many scientists that rushed to confirm Galvani's work was Alessandro Volta. Volta was a physicist. As he probed more and more deeply into Galvani's theory, he saw it beginning to fall apart. Volta soon discovered that the junction of two dissimilar metals could produce electricity without the use of any animal tissue whatever. Volta had little alternative but to challenge Galvani's conclusions. Galvani, of course, defended his work, but Volta's experiments were conclusive and undisputable. Galvani died a few years later, spiritually and financially broken, while his rival was heaped with honors.

Volta's continuation of Galvani's work concluded that electric charges were produced when any two different metals touched. His one wrong conclusion was that it was the contact between the metals that produced the charges. We now know that it is an electro-chemical reaction. Countless experiments eventually produced a list of metals, any of which would produce a positive charge when used with one lower on the list. This led to the development of the voltaic cell, the first basic electric cell.

The voltaic cell consists of a glass container filled with acid. Two pieces of different metal are immersed in the acid, and this develops an electric potential between them. When the electrodes are connected with a piece of wire, current flows. Unlike the instantaneous spark produced by the Leyden jar and other static-electricity devices, the voltaic cell produces a steady current. As

Table 2-1. Electrochemical Potential.

Solution	Sulphuric Acid	Lye	Salt
Metal			
Zinc	0.0	−0.321	?
Lead	0.513	0.318	0.512
Tin	0.513	0.002	0.503
Copper	1.007	0.802	0.809
Silver	1.213	0.958	1.013

When any of the metals listed are immersed in a solution of the given chemicals, the voltage produced between two metals will equal the difference of the indicated numbers. For example, zinc and silver in a solution of lye will produce −0.321 − 0.958, or 1.719 volts.

the current flows, the metal pieces corrode away. Chemical action produces electricity. Table 2-1 shows the electrical relationship of metals.

Figure 2-1 shows a simple voltaic cell, made from a jar filled with sulphuric acid, and two pieces of different metal immersed in it. The metal strips are copper and zinc, and the combination produces approximately one volt.

A voltaic cell that uses acid as the chemical for the reaction is too dangerous for youngsters, but there is one type that is safe to make and handle. It be made from a metal can such as those that you buy vegetables in at the grocery, a few inches of copper tubing, and some vinegar (see Fig. 2-2). This only produces about half a volt, but that is enough to show on a low-cost voltmeter. Similar simple cells are shown in Fig. 2-3.

We cannot do very much with half a volt. There must be some way to get more. Before you can do that, however, let's get a better idea of just what we mean by a volt.

In the last chapter, we were able to produce a difference in electric potential by producing a static charge. When we did this, we made one object either more positive or more negative than the other. The difference in the level of electrical charge produces an electrical pressure, or electromotive force, between the two objects. This is true when one object has a positive charge and the other a negative charge, and it is also true if both objects had either a positive or a negative charge, but one object was charged more heavily than the other.

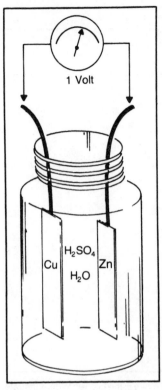

Fig. 2-1. A basic battery. The amount of current it can deliver depends on the area of the plates.

To clarify this, look at Fig. 2-4. The steps shown represent different charge levels. Notice that the step in the middle is called ground. If an object is charged more positively than the ground, it is said to have a positive charge; if it is charged less positively, than ground, it is said to have a negative charge. Generally, all charges are measured with respect to ground, but that is not always the case. Take a second step above ground, for instance, it is twice as positive as the first step. Consequently, the first step has a positive charge with respect to ground, but a negative charge with respect to the second step.

Going in the other direction, the first step below ground has a negative charge with respect to ground, but a positive charge with respect to the second step. It's all a matter of where you set your reference. In the earlier chapter, when we talked about positive and negative, we referred to an otherwise absolutely neutral object. Absolute neutrality is not very easy to find in the real world, so we generally reference the indication of polarity with respect to ground.

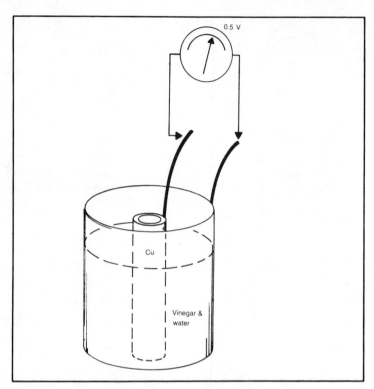

Fig. 2-2. Here is a simple battery that you can make in the kitchen to demonstrate the principles. The copper must not touch the side or bottom of the can.

Fig. 2-3. Here are two antique cells from an old telephone system. Note the glass jar to prevent damage that might result from leaking. (Courtesy Allegheny County Historian.)

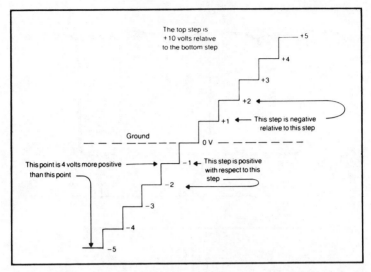

Fig. 2-4. Voltage is an expression of potential difference, and is entirely relative to what you are measuring.

Where does that leave us, so far as our tin-can cells are concerned? It is essential that we understand the foregoing before we begin to build up voltage. The positive and negative terminals are so termed only with reference to one another.

Let us now make a number of tin-can cells. We know that each one has about one half *volt* of electric pressure between the copper and the can. Now, suppose we connect the can of one cell to the copper of another. The two points, being connected together, are of the same potential, thus the can of the first cell is one half volt from the can of the second. The copper of the first cell, being one half volt away from its outside, is one volt away from the can of the other. From this we learn that, if we connect cells together in such a way as to require the current to pass through each in succession, the voltages of the cells will add. The more cells, the higher the voltage, Fig. 2-5. A number of cells connected together is called a *battery*, although common misusage applies this term to a single flashlight cell, calling it a battery even though there is only one cell.

The current supplied by such a battery is no more than one cell alone could supply, the advantage being a gain in voltage. If voltage gain is not needed, but current gain is necessary, the battery can be connected so that the current can divide, each cell supplying current independently. This type of hookup is a parallel connection, Fig. 2-6. While often used in the various devices a battery may supply, a

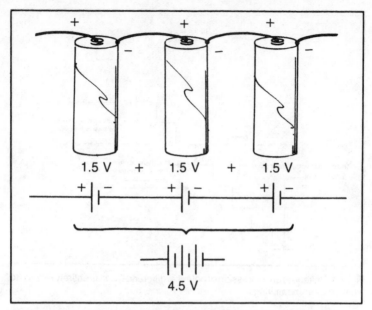

Fig. 2-5. When battery cells are connected in series, their voltages add.

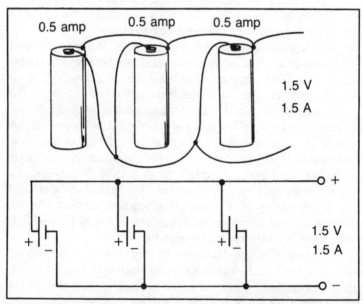

Fig. 2-6. When battery cells are connected in parallel, the total voltage remains the same, but the amount of current that can be delivered is the sum of the current capacities of the individual cells.

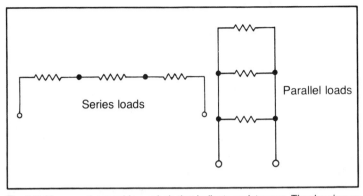

Fig. 2-7. Here are the graphic symbols that indicate resistances. The drawing on the left shows three resistances connected in series; that on the right shows them connected in shunt, or parallel.

parallel battery is rarely, if ever, used. Parallel loads are illustrated in Fig. 2-7.

THE STORAGE CELL

The Leyden jar described in the previous chapter was the first attempt to store electric power. It had its limitations. It could only store a limited amount of energy, and would discharge completely in a very short time. When man learned to produce electric energy from a chemical reaction, it was a big step in the right direction. It now remained to discover an electrochemical reaction that was reversible. That is, a chemical reaction that would produce electricity from the reaction of chemicals, and chemicals that would react the other way when electric current was applied, thus renewing the cell. Let's try an experiment to show a reaction to electricity. You'll need the following equipment:

Two 12-volt batteries
Small aquarium tank
Two water glasses
10 to 12 feet heavy gauge insulated copper wire

Connect the two batteries in series, thereby providing 24 volts. Fill the aquarium tank with water. Add a bit of salt to provide conductivity. Fill the two glasses with water from the aquarium and support them, upside down, a few inches from the bottom of the tank.

Cut two lengths of heavy gauge wire and strip about three

inches from both ends of each length. Connect one wire to the positive battery terminal, and the other wire to the negative terminal.

Place the free ends of the wires into the water, one under each glass as shown in Fig. 2-8.

After current has flowed for some time, you'll see bubbles rising from each wire. As they accumulate in the two glasses, notice that one glass fills twice as fast as the other. The glass with the greater amount contains hydrogen, the other contains oxygen. The passage of current through the water has separated the water into its two elements, two parts of hydrogen to one part oxygen.

What you have done is cause a chemical change by passing electric current through a substance. In this case, you have reversed the previous combining of the elements. Remember that electric current can be produced by the chemical reaction of metals with certain substances. If a combination of chemicals could be found whose reactions could be reversed by forcing current through, we could build an electric cell from which power could be drawn, and that power later replaced by reversing the chemical action. Such a cell is called a storage cell, although it does not actually store electricity. It converts electric power into chemical energy, and afterward delivers that power by reconverting from chemical into electric energy.

There are several combinations of materials that will do this. The most popular, however, uses lead, lead oxide, lead sulphate, and sulfuric acid. These are the basic materials used in the common automative battery, Fig. 2-9.

It is not too difficult to make a storage cell that works on the principle of chemical conversion. CAUTION—this experiment

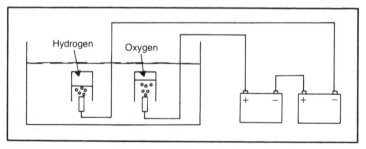

Fig. 2-8. Take a large tank of water. Invert two small jars full of water inside it. Bring insulated wires from the batteries to bare electrodes in the jars. You may have to add some salt to the water to make it conduct. Bubbles will form on the electrodes and go into the jars. One jar will contain hydrogen, the other, oxygen.

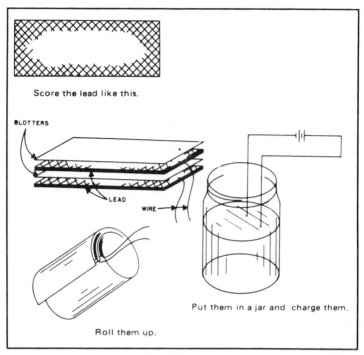

Score the lead like this.

BLOTTERS

LEAD

WIRE

Put them in a jar and charge them.

Roll them up.

Fig. 2-9. A simple storage cell. Be sure the lead strips do not touch one another. Fill the jar with automobile battery acid. Once you have charged the cell, mark the polarity of the charge, and always observe it in future chargings.

should be done by someone who knows how to handle dangerous chemicals. Battery acid (sulphuric acid) can cause severe burns if you get it on your skin, and it will eat holes in almost any cloth, tabletop, or rug that it touches. If you cannot find a chemist, chemistry teacher, or an adult who knows about acid, to help you, pass this experiment up, and take my word for it that it works. Equipment needed includes:

Small glass jar
Sheet lead
Blotting paper
Battery acid

Cut two strips of sheet lead wide enough to fit easily in the jar, and connect a wire to each strip. Scratch a crosshatch pattern on each strip. Roll the two strips with a strip of blotter paper between them, as shown in Fig. 2-9. The strips must not touch each other.

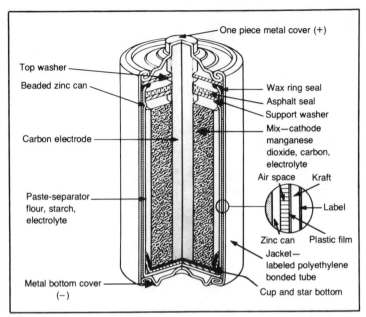

Fig. 2-10. Cutaway of general-purpose flashlight cell. (Courtesy Union Carbide Corp.)

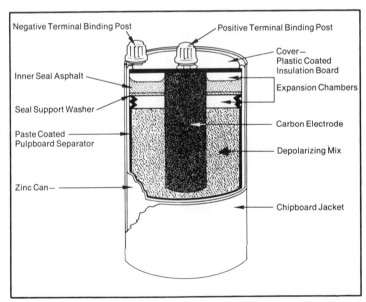

Fig. 2-11. A large #.6 dry cell is just like the flashlight cell inside, only bigger. (Courtesy Union Carbide Corp.)

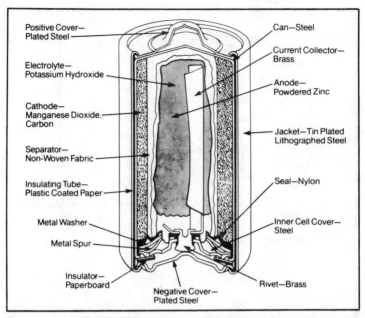

Fig. 2-12. An alkaline flashlight cell. Note how different it is from a zinc-carbon cell. (Courtesy Union Carbide Corp.)

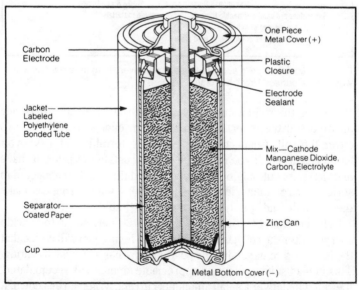

Fig. 2-13. A heavy-duty flashlight cell. It is similar to the standard flashlight cell, and doesn't have quite the capacity of the alkaline battery. (Courtesy Union Carbide Corp.)

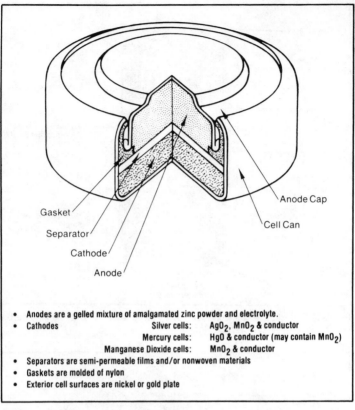

Gasket

Separator

Cathode

Anode

Anode Cap

Cell Can

- Anodes are a gelled mixture of amalgamated zinc powder and electrolyte.
- Cathodes   Silver cells: AgO_2, MnO_2 & conductor
                                      Mercury cells: HgO & conductor (may contain MnO_2)
                         Manganese Dioxide cells: MnO_2 & conductor
- Separators are semi-permeable films and/or nonwoven materials
- Gaskets are molded of nylon
- Exterior cell surfaces are nickel or gold plate

Fig. 2-14. Miniature cells such as these power everything from hearing aids to watches. (Courtesy Union Carbide Corp.)

Place the roll into the jar, and bring the wires out. Mark one positive and the other negative. Pour in enough battery acid to cover the roll. Connect the wires to the terminals of a two-volt battery charger. If you do not have an adjustable charger, or have one that puts out only 6 volts, you'll need three cells connected in series. Charge them slowly (low current); a high charge can make them bubble and splash acid out onto the table.

If the charger is the kind that can be set, set it for trickle charge. After the cell has charged several hours, you will notice that the lead strip connected to the positive wire has a brownish coating. This is lead sulfate. Once an appreciable amount has accumulated, the cell will deliver a small amount of current. When it runs down, it can be recharged.

This cell is intended primarily for experimental purposes. To make a truly efficient cell, a better separator material than blotter paper is needed. Also, the plates should be perforated and pasted—lead sulfate for the positive, lead oxide for the negative.

Other common cells are shown in Figs. 2-10 through 2-14.

VOLTAGE, CURRENT, AND RESISTANCE

This brings us to what is perhaps the most important bit of electric fundamentals you will ever learn. All applications of electric principles depend upon it in one way or another.

Certain materials will allow electric charges to flow freely through them; others will not. Materials that allow the flow of electricity are called conductors; those that do not are called insulators. In recent times, with the development of modern electronics, materials have been discovered that show the characteristics of both. They are called semiconductors.

Electrons flow through a conductor by being forced from atom to atom within the conducting material. The unit of measure for a quantity of electrons is the *coulomb*. When one coulomb per second is flowing through the conductor, it is said to be a current of one *ampere*. The ampere is the unit of measure of electric current. The instrument that measures that current is called an *ammeter*.

The force that moves a current through a conductor is called the electromotive force, or *voltage*. The unit of measure is called the volt. Electromotive force is measured with a *voltmeter*.

Some conductors allow the flow of electric current more easily than others. That is, some conductors have more conductivity than others. The opposition to current flow is called *resistance*. The unit of measure of resistance is called the *ohm*. One ohm is the resistance that will allow a current of one ampere to flow with an electromotive force of one volt.

When current flows through a resistance, power is used up. This power changes into heat, light, or some other form of energy (Fig. 2-15). The unit of measure of electric power is the *watt*. One watt of power is consumed when one ampere of current flows through one ohm.

There is a specific relationship between voltage, current, resistance, and power. It was first defined by a German scientist named Georg S. Ohm. In honor to him, it is called Ohm's law. Ohm's law is the most important single electric formula you will ever learn. The mathematical expression of this law uses the letter E to represent voltage, the letter I to represent current, and the letter R to represent resistance. The basic formula is

$$E = I \times R$$

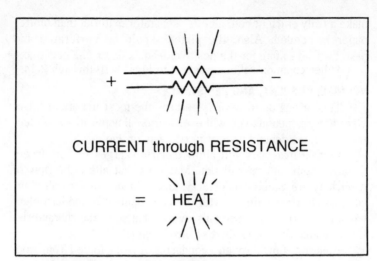

Fig. 2-15. How electric power is dissipated.

Variations of the basic formula allow the calculation of any of the three properties if the other two are known:

$$E = I \times R \quad I = \frac{E}{R} \quad R = \frac{E}{I}$$

In addition to Ohm's law, there is a basic formula to calculate *power*:

$$P = E \times I$$

Combining this formula with Ohm's law, we are able to calculate any of the three basic properties so long as the power is known along with one of the other two:

$$P = E \times I \quad E = \frac{P}{I} \quad I = \frac{P}{E}$$

$$P = E \times \left(\frac{E}{R}\right) = \frac{E^2}{R} \quad P = I \times (I \times R) = I^2 R$$

$$E = \sqrt{P \times R} \quad I = \sqrt{\frac{P}{R}}$$

$$R = \frac{E^2}{P} \quad R = \frac{P}{I^2}$$

If you intend to do any amount of electrical work, either as a hobby or otherwise, you will benefit by committing the basic Ohm's law formulas and the first basic power formula to memory.

THE SIMPLE CIRCUIT

Once a sustained flow of current has been achieved, electricity could be put to work by converting it to light, heat, motion, or some other form of easily-harnessed energy. The device that achieves this conversion, called the load, is connected so that the current flows from the source, through the load, and back again to the source. The round-trip path just described constitutes the basic electric circuit, shown in Fig. 2-16.

Before the nature of electric current was known, it was thought that the electric charges flowed from the positive terminal of the source to the negative. This direction of flow is still referred to as the engineer's *current*. In practice, current flow has been determined to consist of electrons, moving from atom to atom within the conductor. Since a surplus of electrons defines a negative charge, and since the flow must necessarily pass from the surplus to the shortage, we know that the electron flow moves from negative to positive, Fig. 2-17. In solid-state electronic work, another kind of

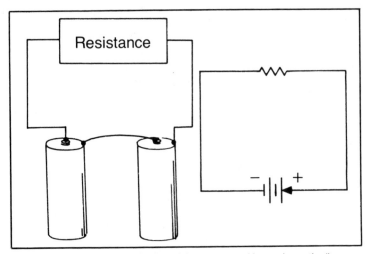

Fig. 2-16. The basic circuit, and how it is represented in a schematic diagram.

41

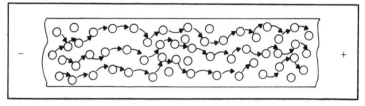

Fig. 2-17. Current flowing through a conductor consists of electrons being forced to migrate from atom to atom. While theoretical *current* is said to flow from plus to minus, the actual *electron* flow is just the opposite.

current flow is defined in which the *absence* of electrons is said to flow in the same direction as engineer's current. This absence of an electron is defined as a *hole*, and will be further explained in a later chapter.

SERIES AND PARALLEL CIRCUITS

Calculating voltage or current in a simple circuit is a straightforward application of Ohm's law. When the circuit contains more than one load, it can become another matter. Loads can be combined into series or parallel circuits. A series circuit is one in which the loads are connected in a chain, like cars in a railroad train. All the current flows through each load, and the voltage applied to each load is in direct proportion to the source voltage as the resistance of the load is to the total circuit resistance.

The total circuit resistance in a series circuit is the sum of all the separate load resistances. For example, if you had three 50-ohm loads, and they were connected in series, Fig. 2-18, the total resistance would be 150 ohms. The current flowing through the loads would be the same as the total delivered by the source, but the voltage across each individual load would be one third of the total source voltage. (This is true only in a series circuit containing three *equal* resistances.)

If the circuit contains unequal load resistances, you must first determine what percentage of the total resistance is under consideration, and multiply that percentage by the source voltage in order to discover the voltage across that part of the load, see Fig. 2-19.

Circuits consisting only of a series combination of loads are quite easy to figure out. When the loads are connected in parallel, it becomes a different ball game. In a series circuit, the current is the same in all loads while the voltage across each load is proportional to the overall load; in a parallel circuit, the opposite is true. The entire source voltage appears across each load and the current is

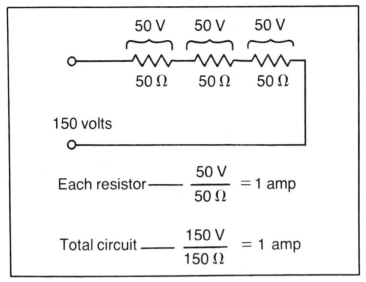

Fig. 2-18. Voltage drop through a series circuit.

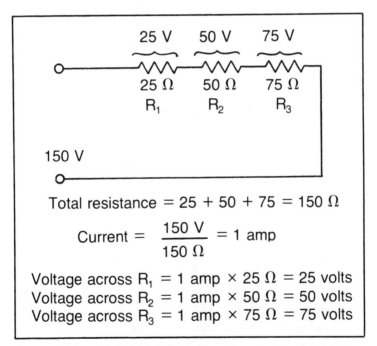

Fig. 2-19. Even if the resistances are unequal, the voltages measured across each of the individual resistors will all add up to the supply voltage.

divided. There are two ways to calculate the overall resistance of a parallel circuit. One is to calculate the currents in the individual loads and then add it up, computing a resistance from the total and the source voltage. The other is to use the reciprocal of resistance, which is known as conductance. Conductance is usually represented by the letter G.

$$G = \frac{1}{R}$$

To calculate the result of parallel resistances, add the conductances:

$$\text{Total } G = G_1 + G_2 + G_3 \ldots$$

or

$$\frac{1}{\text{Total } R} = \frac{1}{R_1} + \frac{1}{R_1} + \frac{1}{R_3} \quad \ldots$$

Where only two or three resistances are concerned, the formula can be simplified this way:

$$\text{Total } R = \frac{R_1 \times R_2}{R_1 + R_2}$$

or

$$R = \frac{R_1 \times R_2 \times R_3}{R_1 + R_2 + R_3}$$

This would work with any number of loads, but over three or so, it becomes a bit cumbersome.

If the loads are equal to each other, a lot of work is eliminated. You simply divide the resistance of one load by the total number of loads. For example, four 100-ohm loads in parallel would have a total resistance of 25 ohms, see Fig. 2-20.

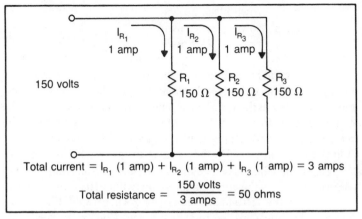

Total current $= I_{R_1}$ (1 amp) $+ I_{R_2}$ (1 amp) $+ I_{R_3}$ (1 amp) $= 3$ amps

Total resistance $= \dfrac{150 \text{ volts}}{3 \text{ amps}} = 50$ ohms

Fig. 2-20. In a parallel circuit, it is the currents that add up.

Series & Parallel Resistors

Let's try an experiment. You'll need a low-cost multimeter capable of measuring resistance, and some carbon resistors (available at any electronic supply store) of 50 ohms, 100 ohms, 150 ohms, and 300 ohms. (Some of these values may only be available as precision resistors.)

Note how the colors on the resistors identify their value. Hold the resistor with the gold or silver band to your right, and observe the following color sequences:

50 ohms: green-black-black

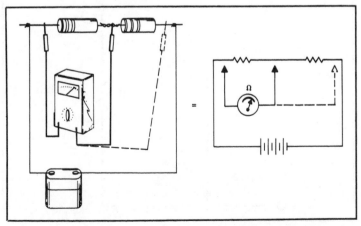

Fig. 2-21. How to measure individual voltage drops in a series circuit.

100 ohms: brown-black-brown
150 ohms: brown-green-brown
300 ohms: orange-black-brown

Connect the 50 and the 100 ohm resistors in *series*. Measure the total resistance, Fig. 2-21. Does it match the calculated resistance according to the formula? Connect a 6-volt battery across the combined resistors, Fig. 2-27. Measure the voltage across the 50-ohm resistor. Measure the voltage across the 100-ohm resistor. Add the two voltages. Do they add up to the battery voltage?

Using the Ohm's-law equations, calculate the current through each resistor. Are the currents the same (within 10 or 20 percent)?

Connect the 150 and 300 ohm resistors in parallel. First calculate and then measure the total resistance. Does your calculation agree?

Basic Doorbell Circuit

Here's an experiment to show a basic doorbell circuit. You will need the following equipment:

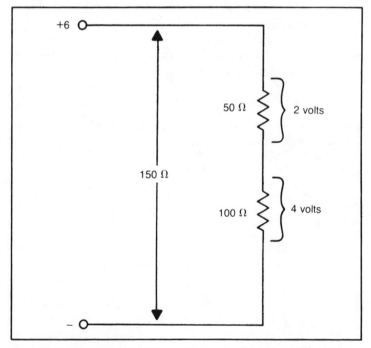

Fig. 2-22. How the resistances in a series circuit relate to the total resistance.

46

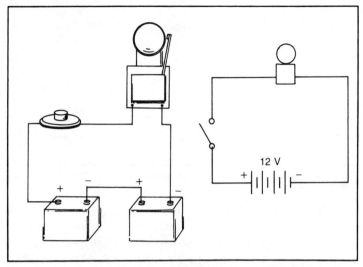

Fig. 2-23. A simple doorbell circuit, and how it is represented in a circuit drawing.

Doorbell, 12-volt
Push button
Single-conductor bell wire (1 roll of 25 feet)
Two 6-volt lantern batteries

Connect the two batteries in *series* (positive of one to negative of the other), as shown in Fig. 2-23. Connect the bell and the button in series, then connect the bell/button combination to the battery.

Try the button and see that the bell rings. This is an example of a simple circuit. Other bells may be connected in parallel with the first bell, and all will ring when the button is pushed. Other buttons can be connected in parallel with the first button and any one button will ring all bells.

Now, let's try a front and rear doorbell circuit. You'll need the battery and the bell as in the previous experiment, plus a buzzer, two buttons, and a bit more wire.

Connect the bell and buzzer in series, as in Fig. 2-24. Connect the wire that is common to the two loads to one battery terminal. Connect the two buttons in series, and connect the wire that is common to the two buttons to the other battery terminal. Connect the second terminal of one button to the second terminal on the bell. Connect the other button to the buzzer.

Push one button and the bell will ring; push the other button and the buzzer will ring.

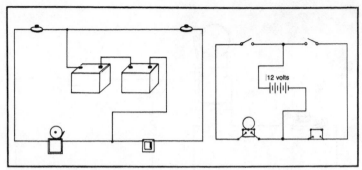

Fig. 2-24. A bell and buzzer for front and back doors, and how they are shown in a circuit drawing.

When electric current flows through a load, power is said to be consumed. This is not entirely true. Electric power is actually converted to some other kind of power—heat, light, motion, etc. All electric devices generate some amount of heat, which is usually dissipated into the air. How much of the power consumed is lost as heat is a measure of the device's efficiency. One hundred percent efficiency can only be realized when the entire output is in the form of heat.

It is, at times, to an advantage to insert a small amount resistance in series with a circuit. If the high-resistance material has a relatively low melting temperature, excessive current will melt it, thus opening the circuit. Such a device, known as a fuse, has protected homes and equipment for over half a century. It is presently being replaced by the circuit breaker, which performs the same job.

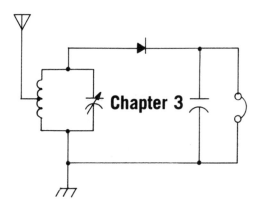

Magnets

No written treatment of basic electrical principles would be complete without some coverage of magnetism. While the initial discoveries associated with this mysterious force were developed independently, magnetism has been found to be inseparable from electrical forces. The two sciences go hand in hand.

Magnetism was originally considered little more than a curious plaything. Today it is an indispensable servant. It is the force that rings our doorbells, turns the blades of our fans, operates the compressors on our refrigerators, produces the sound in our radios, even moves our trains down the track. All these things, however, would not happen except for the link between magnetism and electricity.

While magnetism is the basic force that causes an electric motor to turn, it is useless except where it can be controlled. The control of a magnetic field is accomplished only by the control of electric current. In turn, most electric currents are produced by the influence of powerful magnets. The relationship between electric current and magnetism will be discussed at length in a later chapter. For now, let us be content with examining the force itself.

The first discovery of magnetism is lost in antiquity. It is altogether possible that those mysterious forces came to light in several places simultaneously. Certainly, it is not likely to have happened before the discovery of iron.

The iron age is said to have begun in Asia Minor around the

seventh century B.C. and to have spread over the civilized world in the next five centuries. It is a matter of record that Thales of Miletus studied the curious properties of lodestone (a natural magnet) and amber around 700 B.C.

One of the most interesting accounts of discovery, however, goes back to a shepherd living on a mountainside not far from the site of ancient Troy. Tradition says that, as he idly drew the iron-clad end of his staff along the ground, it seemed to cling of its own accord to an outcropping of black stone. He then noticed bits of the same mineral clinging to the nails of his sandals. The shepherd's name was Magnes, and from that we get such words as magnet and magnetism.

Other legends place the discovery at Magnesia, which was 70 miles or so southeast of Troy. Neither place is very far from Miletus, where we know Thales studied the material. At best, we can say we have a pretty good idea of the region where the discovery was made.

In the early years of discovery, rumors of the mysterious powers of "magnes stone" literally ran wild. Stories about magnetic mountains which would draw the nails from the sides of passing ships terrified sailors. A piece of the material, carried in the pocket, was a cure for a wide variety of illnesses.

Perhaps the most important single discovery was that a piece of the material, suspended by a thread, would always align itself north and south. Thus it became known as lead-stone, later corrupted into lodestone.

Next in importance to the north-seeking properties of lodestone was its ability to impart its power. As man learned to make steel from iron, he also learned that a piece of steel, rubbed with lodestone, would acquire magnetic powers of its own. With this knowledge, mariners could magnetize a fine needle and get precise direction indications, where only approximations were possible with crudely shaped pieces of lodestone.

It was a long, long time before man could understand just what was going on, and even now we barely understand that. Before our understanding could be possible, it was necessary to discover the basic structure of matter.

We now know that all matter is made of atoms, which form molecules. Molecules are always in motion, and the amount of that motion is a direct function of temperature. In a piece of steel, the molecules (which move relatively little due to the hardness of the material) are normally haphazardly oriented. In order for the material to appear magnetized, it is necessary for all of the molecules to be aligned, see Fig. 3-1. Anything that affects the alignment of the molecules, such as overheating or mechanical shock, will weaken or destroy the power of a magnet.

Magnetic materials are iron, nickel, cobalt, or alloys containing one or more of those metals. Most of the better magnets today are made of alnico, an alloy of aluminum, nickel, and cobalt. Generally speaking, the more easily a material is magnetized, the more easily it can be demagnetized.

BASIC MAGNETIC PRINCIPLES

The following simple experiments can be done with just a few magnets:

Two bar magnets
Hammer
Pocket compass
Iron filings
Several small steel bars, not magnetized
Paper
Steel rod (approx. 3 feet long)

Magnetic Attraction and Repulsion

Place the compass on the table, clear of any metal objects. Allow time for the needle to settle down.

Move a bar magnet close to the compass until the needle

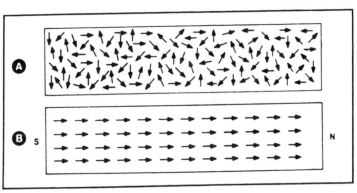

Fig. 3-1. When a piece of metal is not magnetized, the molecules all point in different directions, A. Magnetizing causes the molecules to line up so that all their magnetic fields combine, B.

moves toward the magnet. Note which end of the compass moves toward the magnet. Mark the end of the magnet nearest the compass.

Bring the other bar magnet close to the compass. Try first one end, then the other, until the same end of the compass needle is attracted toward the magnet as before. Mark that magnet also. You now have two bar magnets, each marked with the end that drew the same part of the compass needle. Bring these two ends close together. Note what happens.

Now turn *one* of the bar magnets end for end, and try the same thing. Note the different reaction.

You probably knew what would happen—when you tested the magnets with the compass, you were marking the poles, see Fig. 3-2. First you tried bringing the same poles of two magnets together. Undoubtedly, you noticed they pushed each other away. Then you turned one magnet around—now you were bringing the opposite poles together, and they attracted one another. This demonstrated the primary rule of magnetism. *Like poles repel; unlike poles attract.* (Note the similarity to the behavior of electrical changes.)

INDUCED MAGNETISM

You probably noted that one end of the magnet attracted one end of the compass needle, while the other end of the magnet drew the other end of the compass needle. If you bring a piece of unmagnetized metal close to the compass, you will find that both ends of the metal can attract either end of the compass needle.

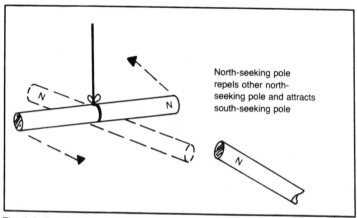

North-seeking pole repels other north-seeking pole and attracts south-seeking pole

Fig. 3-2. The first law of magnetism—like poles repel, unlike poles attract.

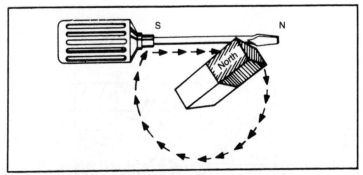

Fig. 3-3. You can magnetize a screwdriver by rubbing it with a magnet. Be sure to always rub in the same direction.

Rub the piece of metal with one end of a bar magnet. Take the magnet away from the metal, and rub the other end of the metal with the other end of the magnet. Do this several times. Now test the metal with the compass, as before. This time you will note that

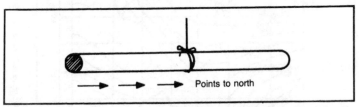

Fig. 3-4. Hang up a bar magnet so that it can swing freely, and it will point north.

when the piece is reversed, the compass needle reverses. It now has "poles." Try picking up some iron filings with it. Note that it now possesses magnetic power, see Fig. 3-3. You have used a magnet to make another magnet. This is called *induced magnetism.*

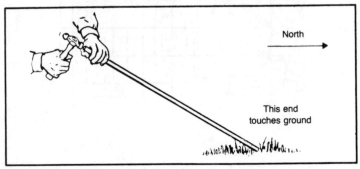

Fig. 3-5. How to magnetize a steel rod from the earth's magnetic field.

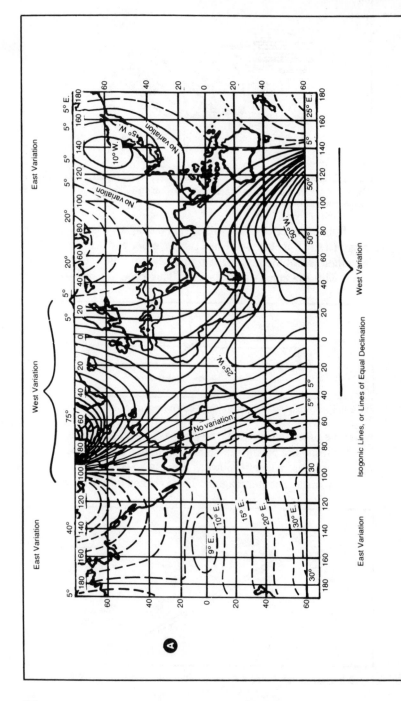

Isogonic Lines, or Lines of Equal Declination

East Variation   West Variation   East Variation
East Variation   West Variation

A

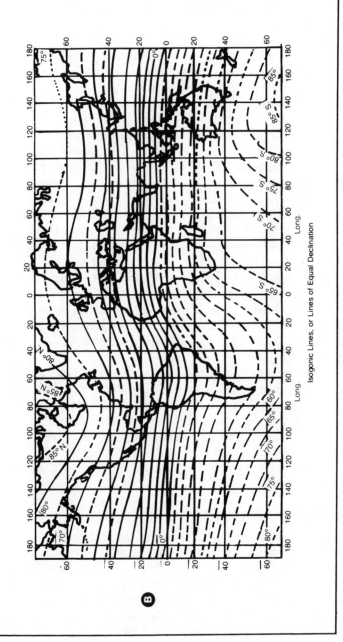

Isogonic Lines, or Lines of Equal Declination

Fig. 3-6. Map A shows deviation of a compass from true north. Map B shows the angle at which a needle will dip downward to the north or south.

55

HOW A COMPASS WORKS

Hang a bar magnet by its center so that it can swing freely. Watch it until it becomes still. Note that it aligns itself in a north/ south direction, see Fig. 3-4.

The ends of a magnet are called *poles*. One is called the north-seeking pole, the other is called the south-seeking pole.

Take a steel bar, about three feet or so in length and determine that it is not magnetized by testing it with a compass. Now rest one end on the ground, and align it in a north/south direction. Strike it several times with a hammer (Fig. 3-5). Now bring it near a compass and test it for magnetism. What do you observe?

You have made a magnet by aligning a bar of metal with the earth's magnetic field. This shows that the earth itself is a magnet, and explains why the compass needle, also a magnet, aligned itself in a north/south direction.

Now take this magnet and see just how strong it is by trying to pick up iron filings. You will find it's pretty weak. Rub it a few times with the bar magnet to make it stronger. If you're lucky, you can pick up a small pin with it.

Heat the bar with a propane torch and again try its magnetic power. Notice any difference? Remember what we said a couple of pages back. When a piece of metal is magnetized, some of the molecules are aligned. Heating the bar set the molecules into violent motion. This upset the alignment and destroyed some of the magnetic power.

Let's go back to that magnet you hung up. You may have noticed that, even though you hung it carefully by the center, it still dips downward at the north end. This downward dipping of a magnetized needle is a natural phenomenon called declination, and increases as one goes further north of the equator. At the earth's magnetic north pole, the bar or needle might point straight down. Figure 3-6 shows the magnetic deviation and declination that affects a compass reading.

TRACING THE MAGNETIC FIELD

Place the two bar magnets used in the previous experiment with the two unlike poles facing one another, but an inch or two apart, and lay a sheet of paper over them. Put some iron filings into an old salt shaker and sprinkle them on the paper covering the magnetic poles. Notice the pattern in which they fall; they almost seem to trace curved lines joining the two magnets.

In fact, they do. Magnetic fields are expressed in terms of *lines*

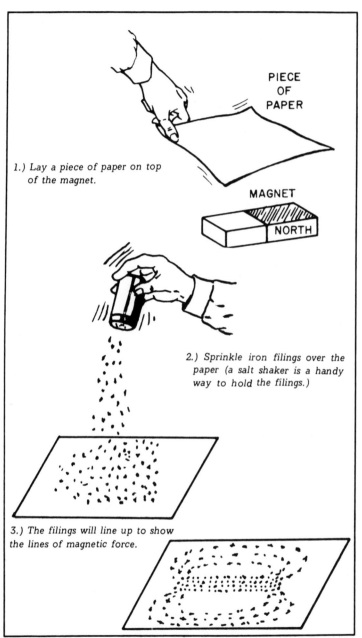

1.) Lay a piece of paper on top of the magnet.

PIECE OF PAPER

MAGNET

NORTH

2.) Sprinkle iron filings over the paper (a salt shaker is a handy way to hold the filings.)

3.) The filings will line up to show the lines of magnetic force.

Fig. 3-7. You can show a magnetic field on a piece of paper by placing it over a magnet and sprinkling iron filings onto it. To make the sketch permanent, coat the paper with glue first.

of force. If you reversed one magnet so that the repelling poles were facing one another, you would see the lines of force from the two magnets curving away from one another instead of connecting, as when you had the attracting poles facing, see Fig. 3-7.

Permanent magnets are used extensively in the home and in industry. They see many applications, ranging from cabinet-door latches to compasses. Their greatest field of application, however, is possible only through the aid of their equally-mysterious companion, electricity.

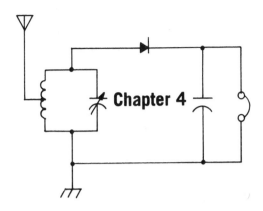

Electromagnetism

For a great many years, the sciences of magnetism and electricity followed parallel courses but were otherwise unconnected with one another. In fact, many scientists believed that neither had anything to do with the other. The mysterious union between them was to remain undiscovered until 1819 when Hans Christian Oersted, a Danish scientist, happened to be lecturing on the heating effects of electricity. A compass was lying on the table close by, and Oersted noticed that the compass needle moved each time he applied current to the wire. Moreover, when Oersted reversed the connections to the battery, the compass moved in the other direction.

Here, then, was the answer: A current of electricity creates a magnetic field. Oersted's discovery was widely published. Davy, in England, noticed that iron filings would cling to a wire when current flowed, and drop away the instant the current ceased. André Ampère, in France, showed that the magnetic field existed all through the circuit—even in the battery cells. The evidence was that the current had to be in motion in order to create a magnetic field. Still, it was the motion of an invisible something, and that wasn't easy to accept.

It was only a matter of time before somebody thought of trying to increase the strength of the magnetic field by winding the wire into a coil of many turns. Now it behaved exactly as a magnet—with north- and south-seeking poles. This was a discovery of monumental importance, but it was only the beginning.

At first, electromagnetism was something of a curiosity. There was good reason for this: insulated wire had not yet been invented. When Ampère made a magnetic coil, he had the wire inside a coiled glass tube. In 1825, Sturgeon discovered that soft iron was very easily magnetized and demagnetized. This paved the way for Joseph Henry to insulate a wire (by hand) with layers of silk and then wind a coil with a great many turns of this insulated wire. One of his early electromagnets lifted a ton.

The electromagnet was finally ready for the world, and one of the first to make use of it was a young American named Samuel Morse. His idea was to use an electromagnet as a signaling device. It wasn't all that new; numerous electromagnetic telegraph devices had been in use in Europe. These, however, were very primitive. Morse's idea simplified the system enough to make it practical.

The idea of an electric telegraph was not by any means a new one. Right from the first realization of the speed with which electric impulses travel through a wire, experimenters had been trying to make an electric communication system. The main problem was how to detect the electric charges when they reached the other end. Many systems had been tried. One used a series of 26 pairs of wires running through glass tubing. These terminated in a tank of water, and the electric signals were read out by the presence of bubbles coming from the wires. That system was annoyingly slow.

In Spain, somebody developed a telegraph line in which the operator at the receiving end *felt* the signals. We need not comment on the sheer pain of a long-winded, high-speed operator!

With the discovery of electromagnetism, the problem seemed to be solved. Intelligence could be instantly taken from the wire. Now the competition was narrowed. One system used 26 wires ending in 26 galvanometers. In England, an enterprising system was devised using only five pair of wires. This system proved worthwhile in helping to catch a dangerous criminal.

When Morse arrived in England to sell his invention, he was greeted with such remarks as "Electromagnetic telegraph? Sure. Just go on down to the railroad station and you'll see one in operation." There was one big difference. Morse's idea used only one pair of wires. It could be adapted into any on/off function. Consequently, Morse is most remembered among the men who pioneered electric communication.

Even though Morse's idea was uniquely his, there were those who questioned his right to the patents. After all, they reasoned, Joseph Henry had invented the electromagnet, and had helped

Morse build the one he used in the telegraph. Morse, you see, was not an electrician; he was an artist. He simply hadn't realized that the wire had to be insulated. Once Henry set him straight, Morse was able to do the rest.

A GALVANOMETER

Some of the early experiments are still frequently duplicated. Here are a couple you can try. You will need the following equipment:

Pocket compass
Dry cell or flashlight cell
Enameled wire (size between No. 20 and No. 30 AWG)

Wind a dozen or so turns of wire in a tight loop around a small jar or other convenient (nonmetallic) form. Partially flatten the coil and secure the turns in place with some tape, see Fig. 4-1.

Mount the coil in a vertical position and place the compass inside it. Position the coil and compass so that the needle comes to rest parallel with the wire (Fig. 4-2).

Connect the coil to the dry cell and observe the compass. Does the compass move? Which direction?

Reverse the wires at the dry cell. Now which way does the compass move?

RIGHT-HAND RULE

Look carefully at the coil and determine the direction of the *current* (positive to negative). Place your right hand over the coil so that the fingers point around the coil in the same direction as the

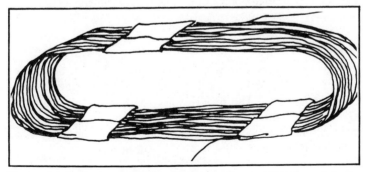

Fig. 4-1. Wind a flat loop and secure it with tape. Be sure it's big enough for the compass to fit inside.

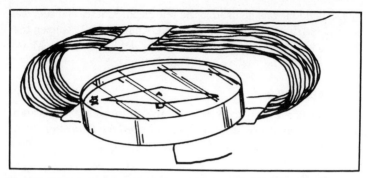

Fig. 4-2. Position the compass within the coil so that the needle lines up with the winding. Then apply power and watch the needle jump!

current. Your thumb will point in the same direction as the north-seeking end of the compass needle.

In demonstrating the behavior of the compass needle with regard to electric current, you have duplicated one of the oldest experiments involving these two factors. You have also made a crude galvanometer—one that was certainly a forerunner to today's meter. Your compass galvanometer will prove to be very sensitive, detecting very small amounts of current.

When you placed your hand over the coil and noted the magnetic polarity, you demonstrated a very important rule in the design of electromagnetic devices. It is commonly known as the *right-hand thumb rule*, and practically every text covering the subject gives this rule in one variation or another.

The most common version of the right-hand thumb rule is expressed this way. Grasp the wire with your fingers of the right hand following the magnetic field (out of the north; into the south), and your thumb will point in the direction of the conventional, or engineer's, current flow (toward the negative), see Fig. 4-3.

PERMEABILITY

Magnetic force travels much better through some substances than others. The ability of a material to contain magnetic force is called *permeability*. The experiments you have performed so far have used coils that had nothing but air in the center. Magnetic materials have a much higher permeability than air. Here is an experiment to demonstrate that. You will need:

Pencil or short piece of dowel
Paper

Scotch tape
Wire (same size as previous experiment)
Large iron nails (2 or 3)
Compass

Roll a strip of paper around a pencil or short piece of dowel, and secure it with some scotch tape. Remove the pencil or dowel from the center, leaving a small, paper tube. Wind several layers of wire around the tube. Leave about 18 inches for connecting to the dry cell, see Fig. 4-4.

Apply power to the coil and see how many nails you can pick up with it. Check with the compass to see how far back the magnet coil will affect the compass needle as the power is turned on and off. Try the right-hand thumb rule to determine the north-seeking pole.

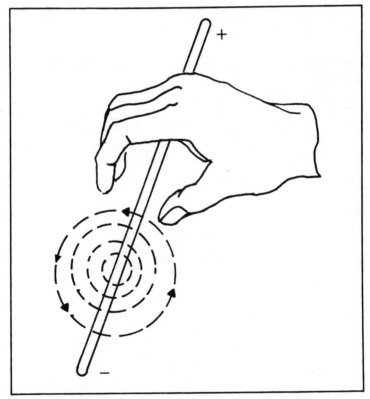

Fig. 4-3. The right-hand thumb rule. Grasp the wire with your thumb pointing toward the negative. Your fingers will go around the wire in the direction of the magnetic field (out of the north, into the south).

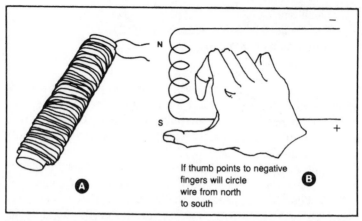

Fig. 4-4. You can use the right-hand thumb rule to determine the poles of an electromagnet, A. Furthermore, it can be slightly modified this way: If you grasp the coil with your fingers pointing in the direction of the electron flow, your thumb will point toward the south pole, B.

Does that end of the coil attract the south-seeking end of the compass needle? See Fig. 4-5.

Insert the nail into the magnetic coil (Fig. 4-6) and again apply power. Can the electromagnet now pick up more nails than before? Does it affect the compass needle from farther away?

You may notice that the nail retains some of the magnetism after the power is removed. This is natural, because the nail is made

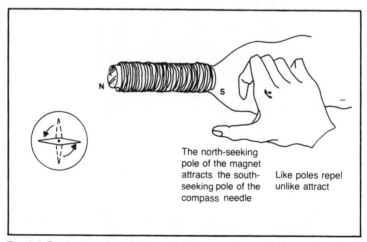

Fig. 4-5. Don't get confused! The end of the compass needle that points toward the south is the one that will point toward the north-seeking pole of the electromagnet.

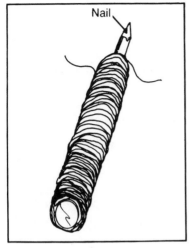

Fig. 4-6. The electromagnet becomes much more powerful with an iron or steel core.

of steel, not soft iron. If you try a soft iron rod, you may notice still more magnetic power and less residual magnetism.

SIMPLE SOLENOID

The difference in permeability between air and a magnetic metal is so great that the coil will try, with considerable force, to draw the nail into a position where it can contain the maximum concentration of magnetic force. This position is, of course, in the

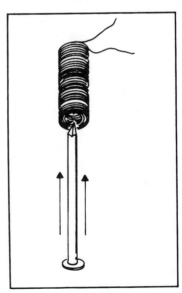

Fig. 4-7. The magnetic field seeks the path of least resistance. Consequently, a hollow coil will draw the nail right up inside.

center of the coil. Such a device is called a *solenoid*. Try the following equipment:

Position the nail so that the point is just about to enter the coil. Energize the coil. What happens? Reposition the nail. This time hold on to the nail while somebody else energizes the coil, see Fig. 4-7. Note how much force is exerted.

The discovery of electromagnetism was a major step toward the use of electricity for communication, as mentioned at the start of the chapter. Now let's try to retrace some of the early steps that led to the present state of the art.

THE TELEGRAPH

Morse's electric telegraph did not, as one might think, originally work by sound. The electromagnet moved a pen over a moving

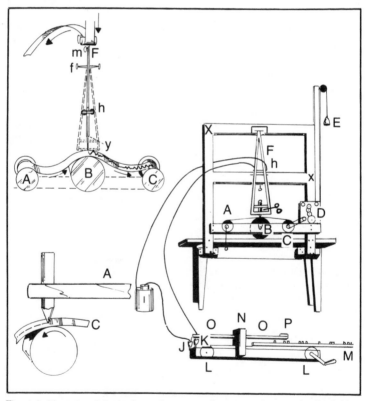

Fig. 4-8. Morse's original telegraph was nowhere near the way the final thing ended up. Morse didn't realize the operator would find it easier to copy the letters by sound, so he had a magnetically-operated pen mark a moving paper strip.

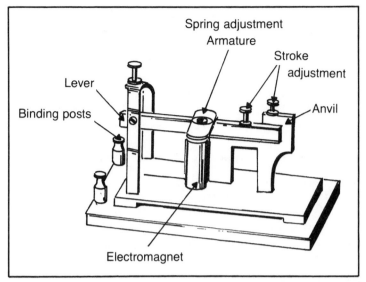

Fig. 4-9. The operator's ability to copy by ear made the invention much simpler, as shown here.

strip of paper that recorded the variations in the electric current (Fig. 4-8). The marks on the paper were later read and interpreted as letters. The movement of the pen made a plainly audible sound which operators quickly learned to interpret, reading the message as fast as it came over the wire. In time, the pen and paper were done away with, and the Morse telegraph *sounder* (Fig. 4-9) became

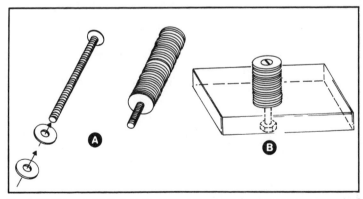

Fig. 4-10. Magnet details for telegraph sounder (A). Be sure to leave enough of the bolt sticking out to mount it on the base. Counterbore the mounting hole from the bottom of the base enough to clear the mounting nut (B). Do not over-tighten the nut or you will damage the coil.

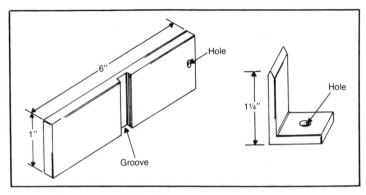

Fig. 4-11. Rocker arm and mount.

the mainstay of electric communication until the invention of the telephone.

The telegraph sounder consisted of an electromagnet that pulled a movable arm against a stop, thus making a clearly audible click. When the magnet released, the arm was pulled back by a spring against another stop. The sound consisted of two clicks, and the space between the two determined whether the signal element was a dot or a dash.

A simplified telegraph sounder can easily be made with an electromagnet, a short steel bar, a spring, and a few miscellaneous scraps.

The sounder may be mounted on a wood base or, for louder sound, it may be mounted on a small wood box. The heart of the sounder is an electromagnet. Use a bolt about 2 inches long for the

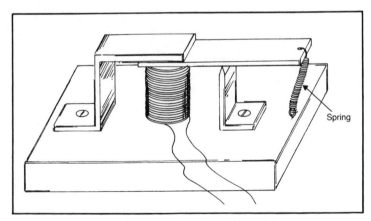

Fig. 4-12. The completed sounder. It's crude, but it works.

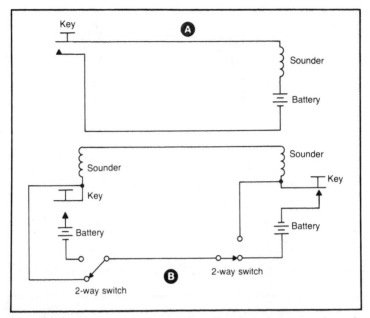

Fig. 4-13. A one-way (A) and a two-way (B) telegraph system, shown in schematic-diagram form.

core. Cut a pair of paper disks for the ends and place them on the bolt, leaving about half an inch of thread for mounting, see Fig. 4-10.

Wind the magnet with No. 28 enameled wire to an overall thickness of about 1 inch. Leave about six inches of wire for connections. Mount the magnet on the base as shown in Fig. 4-11.

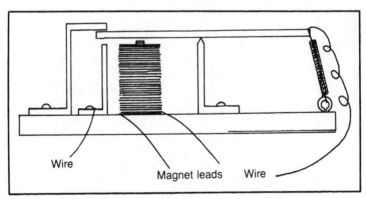

Fig. 4-14. Adding one piece converts your sounder into a relay. When the coil is energized, the circuit is closed between the armature (rocker arm) and the new bracket.

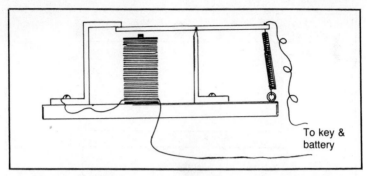

Fig. 4-15. When you direct the circuit through the armature and the backstop, it's a buzzer.

Fashion two pieces of flat steel bar stock as shown in Fig. 4-12. Bend a third piece of bar stock into a Z shape to act as a stop for the sounder. Assemble the three onto the base. Attach a coil spring to the end of the sounder bar, and to the base, as shown in Fig. 4-13.

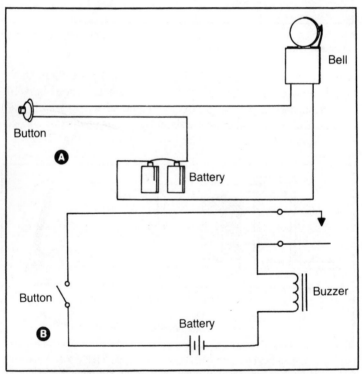

Fig. 4-16. Pictorial (A) and schematic (B) of a simple doorbell circuit.

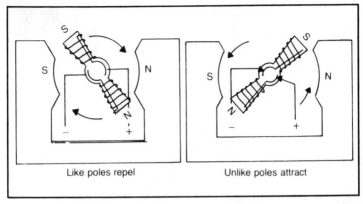

Like poles repel   Unlike poles attract

Fig. 4-17. Basic motor. Note that the power reaches the armature magnet through brushes. The polarity of the magnet is made to reverse as the armature turns.

You can make a key if you really want to, but it's much easier to buy one. They are available at low cost at any number of electronic supply stores.

Wire the magnet, key, and sounder as shown in the diagram, Fig. 4-14. A second sounder and key, all wired in series as shown, will result in a two-way telegraph system.

To operate your two-way system, one key must be closed while the other is operating. Operating either key while the other is closed will cause both sounders to operate, enabling the operator to monitor what he is sending. If the other operator wants to interrupt, he simply opens his key.

The code used by telegraph operators was a little different from the international radiotelegraph code used today. The two codes are shown in Table 4-1. You will notice there are just enough differences to make it confusing.

The thing that really helped Morse's invention was the electromagnetic relay. This allowed one electric circuit to control another, thereby compensating for line losses. In fact, the sounder was developed from the early relays which made so much noise the operators could read the messages before looking at the lines on the paper.

A Relay

A sounder, as made in the previous paragraphs, can easily be converted into a relay. All one need do is add an upright contact piece as shown in Fig. 4-15. Be sure the wire connected to the

71

Table 4-1. The Radiotelegraph Code Compared with Morse's Original Code.

	Land Telegraph	Radio Telegraph
A	· —	· —
B	— · · ·	— · · ·
C	· · ·	— · — ·
D	— · ·	— · ·
E	·	·
F	· — ·	· · — ·
G	— — ·	— — ·
H	· · · ·	· · · ·
I	· ·	· ·
J	— · — ·	· — — —
K	— · —	— · —
L	———	· — · ·
M	— —	— —
N	— ·	— ·
O	· ·	— — —
P	· · · · ·	· — — ·
Q	· · — ·	— — · —
R	· · ·	· — ·
S	· · ·	· · ·
T	—	—
U	· · —	· · —
V	· · · —	· · · —
W	· — —	· — —
X	· — · ·	— · · —
Y	· · · ·	— · — —
Z	· · · ·	— — · ·
1	· — ·	· — — — —
2	· · — ·	· · — — —
3	· · · — ·	· · · — —
4	· · · · —	· · · · —
5	— — —	· · · · ·
6	· · · · · ·	— · · · ·
7	— — · ·	— — · · ·
8	— · · · ·	— — — · ·
9	— · · —	— — — — ·
Zero	— — —	— — — — —
.	· · — — · ·	· — · — · —
,	· — — ·	— — · · — —
?	— · · — ·	· · — — · ·

72

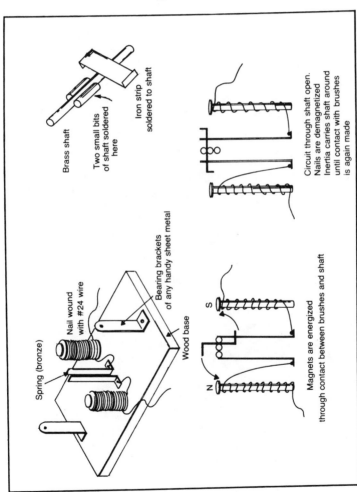

Brass shaft

Two small bits
of shaft soldered
here

Iron strip
soldered to shaft

Nail wound
with #24 wire

Spring (bronze)

Bearing brackets
of any handy sheet metal

Wood base

S

N

Magnets are energized
through contact between brushes and shaft

Circuit through shaft open.
Nails are demagnetized
Inertia carries shaft around
until contact with brushes
is again made

Fig. 4-18. Details of toy electric motor.

73

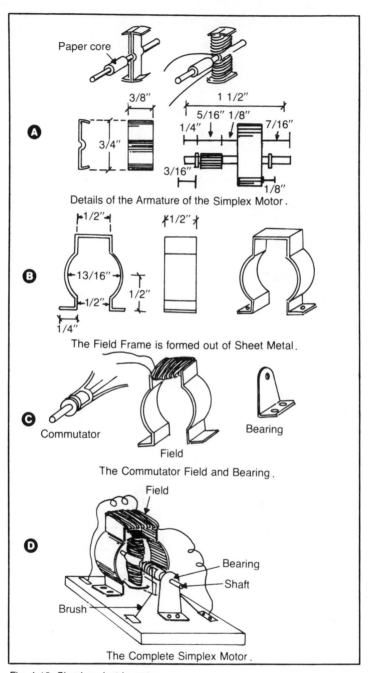

A Paper core

3/8″   1 1/2″

3/4″   5/16″ 1/8″

1/4″   7/16″

3/16″   1/8″

Details of the Armature of the Simplex Motor.

B 1/2″   1/2″

13/16″ 1/2″

1/2″

1/4″

The Field Frame is formed out of Sheet Metal.

C Commutator   Bearing

Field

The Commutator Field and Bearing.

D Field

Bearing

Shaft

Brush

The Complete Simplex Motor.

Fig. 4-19. Simplex electric motor.

74

armature, or movable piece, is flexible enough for the armature to move freely.

If you use the Z-shaped stop as a relay contact, and then connect it in series with the coil, the sounder becomes a buzzer. As the coil is energized and pulls the armature down, the circuit is opened and the coil demagnetizes. The armature then springs back, closing the circuit, and the whole cycle repeats, see Fig. 4-16. The pitch of the buzzer is determined by the power of the spring and the magnet coil. If the stop bracket is adjusted up or down, the pitch varies slightly. Do not try to adjust the bracket with your bare hands as the coil tends to produce a voltage that can give you an unpleasant tingle. This will be explained in a later chapter.

If you are studying for a ham-radio license, the buzzer can be a handy aid to learning the telegraph code. Be sure to use the International Morse code.

The door buzzer or bell is a simple kind of telegraph, in which the button outside the door is the key, and the bell or buzzer inside the house is the sounder.

The electric motor is a trick use of electromagnets in which the polarity of magnets are made to reverse thereby turning a shaft (Fig. 4-17). This is done in the old dc motors by means of brushes which deliver power to electromagnets on the armature, shown in Figs. 4-18 and 4-19.

The next chapter gives plans for a basic dc motor, and for a novel toy motor that doesn't reverse the magnet poles, but works by rapidly switching the magnets on and off.

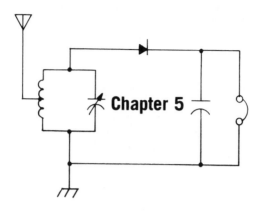

The Mysterious Alternating Current

Up to the very end of the last chapter, we have dealt primarily with stationary charges or with steady-state currents. This was indeed the state of the art in the early years. However, with the invention of the electric motor, and subsequently the generator, a new phenomenon entered the scene. This newcomer is known as alternating current. By alternating current, we mean electric current that alternately reverses direction a number of times a second. Such a current first existed in the armature of electric motors. As ac became more and more prevalent in the electrical sciences, it became more and more a puzzle. For one thing, it didn't seem to obey all of the Ohm's law rules, which seemed to be inflexible laws of nature. It wasn't until such geniuses as Steinmetz, Tesla, and others figured out just what was going on that alternating current became at all useful.

ELECTROMAGNETIC INDUCTION

It seems a peculiar way to start, but the strange behavior of alternating current devices can best be understood by starting with direct current devices. Remember the experiment with the electric buzzer? You were warned that you could get a tingle if you touched the coil while the buzzer was operating. Now let's see just what was happening. Obtain the following equipment:

12-volt battery
Two pounds enameled wire (No. 30 AWG)

Large bolt, about ½ inch by 3 inches, with nut
Two cardboard disks, 3 inches wide, with a hole in center large
    enough for the bolt to fit tightly.

Fashion a spool by fitting the cardboard disks onto the bolt.
(Younger experimenters may need supervision or help here.) Wind
the entire length of wire onto the bolt to make a large electromagnet. If you use two 1-pound spools, be sure to strip the ends and
splice the two together (see Fig. 5-1). Leave about a foot of wire at
either end for connecting to the battery.

Connect the wires to the battery. With fingers touching the
bare part of each wire, remove the wires from the battery terminals
(Fig. 5-2). If the experiment has been done correctly, you should
feel a very definite kick of voltage. You may even see a spark jump
as you remove the wire from the battery.

What happened? Let's see if we can figure it out. You made a
large electromagnet, and connected it to the battery. This created a
magnetic field of considerable intensity around the magnet. When
you broke the connection, the magnetic field instantly collapsed.

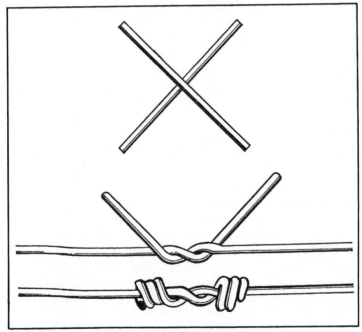

Fig. 5-1. The approved way to join two pieces of wire. Make sure it's mechanically
tight, then solder it.

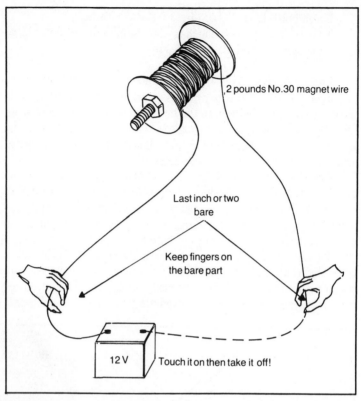

2 pounds No.30 magnet wire

Last inch or two bare

Keep fingers on the bare part

12 V

Touch it on then take it off!

Fig. 5-2. The induced back voltage is a function of amperes and turns. If you pack enough wire into this inductor, you can get a snappy little jolt when the circuit is broken.

There are a great many turns of wire around the bolt. As the magnetic field collapsed, the fast-moving lines of force cut the turns of wire in the coil. This generated a momentary voltage in the coil, in the direction opposite the battery voltage. You were holding the bare ends of the wire, and with the wires disconnected from the battery, there was no place for the charge to drain, except through you. It was the same sensation Cuneus felt when he discharged the terrible jar at Leyden, Holland (see Chapter 1).

Briefly stated, a moving magnetic field induces an electromotive force when it cuts a wire. The intensity of the magnetic field depends on the amount of current, the number of turns, and the permeability of the core. With two pounds of No. 30 wire, you had many turns on that bolt. The 12-volt battery produced a husky current through the wire. Consequently, there was a fairly heavy magnetic field. When the field collapsed, cutting through all those turns of wire, the induced voltage was enough for you to feel.

MOTORS

In the preceding chapter, I deliberately held off from getting too detailed in the explanation of the operation of electric motors for two reasons. First, their operation uses a moving magnetic field which, after all, is what this chapter is all about. Second, the principles of a motor leads directly into electric generation, which works on the principle you have just demonstrated with the preceding experiment.

Let us now take a closer look at the operation of a simple electric motor. Unlike most motors, this one had no wiring in the armature.

When the rotor blades are up and down, the contacts on the shaft touch the brushes, see Fig. 5-3. This energizes the magnets,

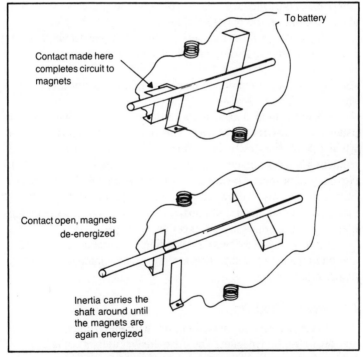

To battery

Contact made here completes circuit to magnets

Contact open, magnets de-energized

Inertia carries the shaft around until the magnets are again energized

Fig. 5-3. All this motor does is turn two magnets on and off. If the armature is a magnet, constantly reversing direction, like the more complex motor, it would have more power.

pulling the rotor around in whichever direction you had started it moving. As the rotor blades are pulled toward the magnets, the contacts open the circuit and the magnets de-energize. Inertia carries the blades around until they are again vertical. At that point, contact is restored between the brushes via the shaft, and the magnets again pull the rotor blades. This is, of course, a toy. It has very little power. It does, however, demonstrate that electromagnets can be used to turn a shaft, the first step toward harnessing electrical power.

The toy motor has very little power because it is, after all, nothing more than a couple of electromagnets pulling against the metal blades. Greater power could have been obtained if the electromagnets pulled against magnets. This is the case in the more complex motor. The electromagnets are in the armature, and pull against two stationary magnets. As the armature turns, the action of the brushes on the commutator pieces reverses the polarity of the magnets in the armature just as they pass the stationary magnets. This causes the armature magnets to repel the stators while pulling at the stators on the opposite sides.

The stator (or field) magnets can be either permanent magnets or electromagnets. If they are electromagnets, they can be connected in series or in parallel (shunt) with the armature brushes, see Fig. 5-4. The connection makes a significant difference in the operation of the motor.

When the field is in series with the armature, its windings are usually made of very heavy wire. The motor draws a heavier current when it is turning slowly, but it turns with greater force (torque). A shunt motor has a thinner gauge wire for the field, but since the field current is constant, the torque is constant at nearly all speeds. Series motors are used when the motor must start under a heavy load with a great amount of inertia. The traction motors in diesel-electric train engines are an excellent example.

Some motors have two field windings, one connected in series, and one connected shunt. These motors offer some of the advantages of both types.

A SIMPLE GENERATOR

So far, we have discussed motors with electromagnets for the stationary, or field magnet. The inexpensive motors available in the hobby shops have permanent magnets for the field. These perform in the same way as shuntwound motors, and can be used for the

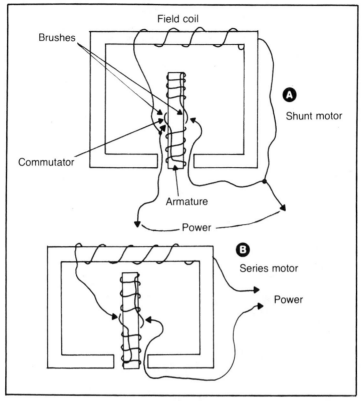

Fig. 5-4. Difference between shunt (A) and series motor (B). Note that while the current in the field is direct current, that in the armature keeps reversing direction.

following demonstration. You will also need a sensitive dc galvanometer.

Connect the galvanometer to the motor terminals (Fig. 5-5). Spin the motor shaft with your fingers and observe the galvanometer. What happens? Can you guess why?

If the experiment was done correctly, the needle of the galvanometer moved when you spun the motor shaft. Remember that the motor consists of coils of wire in a magnetic field. It is easy enough to understand the principle of the electromagnets pulling the shaft around, but there is more to it than that. Any wire carrying a current within a magnetic field produces motion as the current varies. The reverse is true: If a wire is moved within a magnetic field, an electric voltage is produced. If there is a complete circuit,

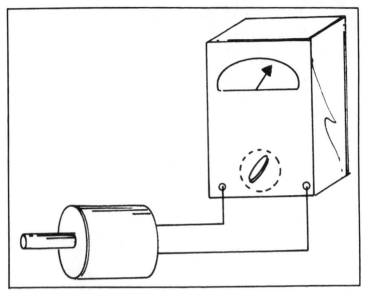

Fig. 5-5. Just as electricity can produce motion with the aid of magnets, motion can, aided by magnets, produce electricity. This is the basis for the operation of every electric generator.

current flows. The amount of the current is dependent on the intensity of the field and the speed of motion. If the circuit has very low resistance, a large amount of force is needed to move the wire.

If you were to take the toy motor used in the preceding experiment and connect it to a flashlight lamp instead of the galvanometer, you might be disappointed. It will take very rapid spinning of the shaft to light the lamp. Just imagine the energy needed to operate an electric stove!

Whether it is operating a flashlight lamp or lighting the whole city, the principle is the same. Wire, often very heavy gauge wire, is turned, by massive stream turbines, within the field of tremendously powerful magnets. That is what produces the electricity that lights your whole neighborhood.

There is another difference between the electricity from a battery and that which is in the wiring of your home, other than voltage and available current. That difference is that the battery current flows continuously in one direction, while the domestic supply reverses its direction 120 times per second. Current that constantly flows in one direction is called *direct current*. Current that is constantly reversing its direction is called *alternating current*.

There are a number of reasons why alternating current is preferred in domestic power supplies: It is easier to generate; it is easier to transmit over long distances; and it can be easily transformed to a higher or lower voltage.

Before we examine the behavior of alternating current, there are a few terms we must understand. These terms apply to the properties and behavior of alternating current.

SINE WAVES

In its purest form, alternating current varies in a sinusoidal manner. By that we mean that, if the variations of voltage or current were plotted with respect to time, they would produce what is called a sine curve on the graph paper, shown in Fig. 5-6.

Note that the time is marked in degrees, not in seconds or in any of the conventional units of time measurement. There is a good reason for this. The *shape* of the wave is the same no matter how high or low the frequency. The measurement in degrees refers to the rotation of the armature of the generator. When the generator has made one complete revolution, the voltage or current has reversed its direction twice and is back to the exact condition in which it started. This condition is referred to as one *cycle*. A cycle consists of *360 degrees*.

The rate at which alternating current varies has long been expressed in cycles per second. One cycle per second is called one *hertz* (Hz). The domestic power supply in the United States alternates at the rate of 60 hertz, or 60 cycles per second. In some countries a frequency of 50 hertz is used. Aircraft, where light-

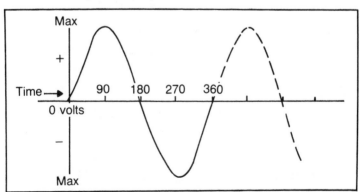

Fig. 5-6. An alternating-current sine wave. As time progresses, the voltage rises from zero to maximum, falls back through zero to maximum in the opposite direction, then repeats.

Table 5-1. Sine of Common Angles.

Angle	Sine of Angle
0°, 180°, 360°	0
10°, 170°, 190°, 350°	0.173
20°, 160°, 200°, 340°	0.342
30°, 150°, 210°, 330°	0.500
40°, 140°, 220°, 320°	0.643
45°, 135°, 225°, 315°	0.707 ◄── Very important electric constant
50°, 130°, 230°, 310°	0.766
60°, 120°, 240°, 300°	0.866
70°, 110°, 250°, 290°	0.939
80°, 100°, 260°, 280°	0.985
90°, 270°	1.000

weight equipment is essential, often use a frequency of 400 hertz. For most applications, however, a frequency of 60 hertz may be assumed.

A sine wave is so called because the voltage or current varies proportionally to the sine of the number of degrees. For example, the sine of 90 degrees is 1. 90 degrees after a cycle has started, the voltage or current reaches its peak value. Another 90 degrees, 180 degrees from the beginning, and the curve is crossing the zero line. The sine of 180 degrees is 0. If we examine the instantaneous voltage at, say 30 degrees, where the sine is 0.5, we find that it has reached a level of 0.5 times the peak, see Table 5-1. The relationship holds true for any instant in the alternating current cycle, as long as the wave is pure and undistorted.

With this basic understanding of what is happening, we can now examine alternating current further. Suppose we energize an electromagnet with alternating current. We can do this using the electromagnet that was built for the preceding experiment, and a bell transformer available at any hardware store. We know that the voltage, and consequently the current in the magnet, is varying at a 60-hertz, sinusoidal rate. Consequently, the magnetic field is varying at the same rate.

TRANSFORMERS

Remember that a wire moving through a magnetic field will produce an electric voltage. This will also happen if the magnetic field is moving. Let's prove it. Obtain this equipment:

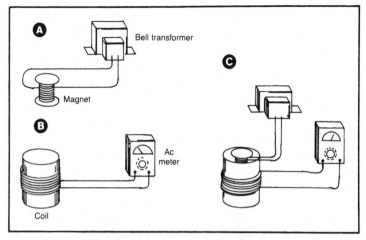

Fig. 5-7. The two coils are not connected electrically, but current in one can induce current in the other. Can you figure out why?

Magnet used in previous experiment
Paper tube that will fit over the magnet
About ½-pound spool of No. 30 AWG enameled wire
Bell transformer
Ac voltmeter

Wind the wire around the paper tube to produce a hollow coil that will fit over the magnet, see Fig. 5-7. Leave about a foot of wire at each end for connections. Connect the hollow coil to the ac voltmeter. Energize the magnet by connecting it to the bell transformer. Plug the bell transformer into an electric outlet. Slip the hollow coil over the magnet and observe the meter, Fig. 5-7C. What happens?

The magnetic field was varying at the rate of the domestic power supply (60 Hz) and this generated an alternating voltage in the hollow coil when it was slipped over the magnet. This kind of device is called a *transformer*. A transformer can isolate the source of power from the load, and can step the voltage up or down, as desired.

The changing of voltage by a transformer is a function of the number of turns of wire in the two windings. The winding that is connected to the source is called the *primary*; the winding connected to the load is called the *secondary*. A transformer may have any number of primary or secondary windings, and the functions of the two can even, in some cases, be interchanged.

If the secondary winding has twice the number of turns as the primary, the secondary voltage is twice that applied to the primary, but the maximum current that can be delivered is reduced. In this case, the primary will carry twice the current as the secondary. Consequently the power in the two windings will be the same (minus a small percentage for losses, which depends on the efficiency of the device). The voltage ratio between the two windings is the same as the turns ratio. The current ratio is the inverse of the turns ratio, see Fig. 5-8.

Alternating and direct current can be run through the same wire at the same time, and afterward isolated by the particular equipment each is supposed to operate. When this is done, the dc level varies at the rate of the ac alternation. This is called direct current with an alternating-current component (Fig. 5-9). Methods to isolate one from the other will be discussed later.

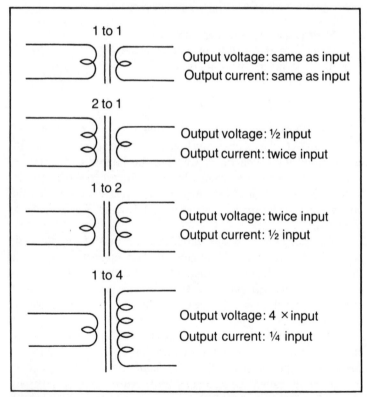

Fig. 5-8. How the ratio of turns between the two windings affects the voltage and current.

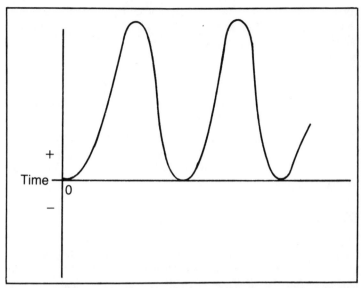

Fig. 5-9. This wave moves in one direction only, but it still varies similarly to a sine wave. In a transformer, it will have an effect similar to alternating current. It's called dc with an ac component.

The current in the coil of a doorbell or buzzer can be described as direct current with an ac component. The main difference is that the alternating component is a square wave shape instead of a sine wave. It can still be made to function as if it were alternating

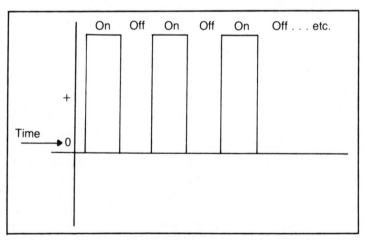

Fig. 5-10. A string of on - off functions like this will pass through a transformer if they're fast enough. High-speed pulses are the workhorses of computers.

current. The magnetic field expands and collapses in the same way as if it were true alternating current, except that the polarity does not reverse, see Fig. 5-10. It will still induce an alternating current in the secondary of a transformer.

This principle is applied in most automobile engines. The ignition coil in an automobile is in fact a special kind of a transformer. It has only one winding, and is called an autotransformer. The breaker points in the system interrupt the direct current from the battery to provide the square-wave component. Battery power is connected across a small portion of a coil that has a great many turns. The step-up principle is the same as if there were two windings, see Fig. 5-11.

ELECTRIC SPARKS

When the automobile was a new invention, the ignition coil was made much more powerful than was needed. This was especially true with the Model T Ford, which had a coil so potent that it's still famous to this day. It was used in all manner of practical jokes, and even used to make some of the first Amateur Radio transmitters. There are still some of them around, and modern-day replicas can be obtained through antique-auto dealers. The following experiments can demonstrate the action of the ignition and spark coils. These experiments are recommended for adults, or under adult supervision. You will need:

Automobile ignition coil
Battery, 12 volt
Wire
Low-voltage buzzer
Dowel or plastic rod, ¼ inch by two feet
12-volt bell transformer

Wire the coil to the battery as shown in Fig. 5-12. Leave one wire to the battery disconnected. Fasten the high-voltage wire to the end of the dowel as shown. Position the high-voltage wire about ¼-inch from the ground wire. Watch the gap between the two wires, and momentarily touch the battery lead to the battery terminal. A spark should jump the gap between the high-voltage lead and the ground lead. If you see no spark, move the wires closer together and repeat the experiment. You will note that the spark jumps most readily when the battery lead is removed. There may be no spark at all when contact is made.

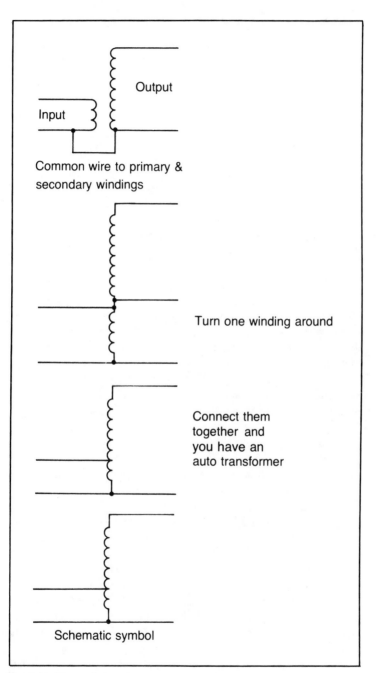

Fig. 5-11. The evolution of the auto transformer.

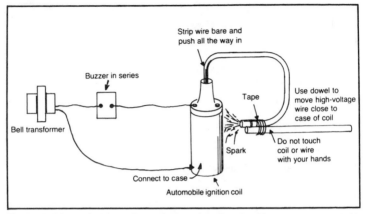

Fig. 5-12. An auto ignition coil was designed to operate with a square wave, not a sine wave. The buzzer in series makes operation much more efficient. As long as the buzzer's there, it will operate also from a battery.

When contact was made, the inductive effect of the coil momentarily hindered the flow of current, causing the current to come gradually (in a few milliseconds) from zero to maximum. Thus, the magnetic field in the autotransformer moved slowly from zero to full strength. However, when the current was interrupted, it stopped instantly. The magnetic field moved quickly from maximum to zero. The faster movement of the magnetic field in collapsing resulted in a higher voltage being generated. Thus, a spark jumped more readily when the circuit was broken.

Now connect the coil to the secondary of the bell transformer. *Be sure the power to the transformer is off.* When connections have been made, apply power to the transformer and, using the dowel as a handle, move the high-voltage wire to the ground lead until a spark is observed. The gap jumped by the ac spark will be less than that jumped by the dc spark, but the spark will jump continuously as long as power is applied, not just when power is interrupted. This is because the magnetic field is continuously in motion.

Connect a low-voltage buzzer *in series* with the primary connections of the coil as shown. Repeat the previous experiment using the battery/buzzer instead of the bell transformer. You will notice that there is a continuous spark, but that it jumps a much larger gap than with the bell transformer supplying the ignition coil. The reason is that an ignition coil is designed to operate with a square wave, not the sine wave supplied by the transformer. With a square wave applied to the coil, the magnetic field moves much faster than with a sine wave. The higher voltage is produced with the square wave, thereby producing the bigger spark.

Electronic ignition systems use a small electronic device to produce a square wave for the ignition coil instead of using breaker points.

The Climbing Spark

Here is a novelty demonstration using a climbing spark. You'll need this equipment:

Ignition coil and buzzer
Wood base
Two feet of rigid steel wire, (about 1/16-inch diameter)

Cut the steel wire into two equal lengths. Shape and mount them as shown in Fig. 5-13. Connect one to the high-voltage lead of the coil, and the other to the ground lead. Apply power to the coil. A spark will instantly jump the gap at the bottom. Then the heat of the spark will cause it to rise until it reaches a point where the gap is too wide to sustain it. The spark will then be broken at the top, and a new spark will immediately begin at the bottom.

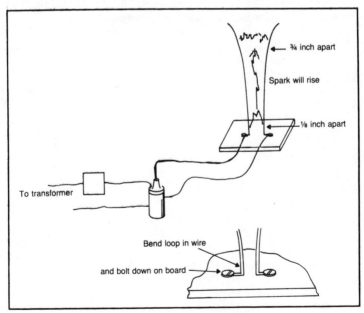

Fig. 5-13. The climbing spark—a mad scientist's delight!

NOTE: Any experiment that produces an electric spark can cause radio or TV interference. Be considerate of your neighbors and do not use these experiments often.

Alternating current can be isolated from direct current in the same wire by means of either or both of two simple devices. Alternating current passing through a coil will be impeded by the action of the magnetic field as it induces voltage in opposition to the applied voltage. By connecting a coil in series with the line, you allow direct current to pass through easily while resisting the alternating-current component. A coil opposes the flow of alternating current, and the amount of opposition is greater at high frequencies. This opposition is called inductive reactance.

On the other hand, if a capacitor is connected in series with the line, the result is opposite to that of a coil. A capacitor consists of two pieces of conductive material separated by an insulator. As the voltage on one side of the capacitor varies, electrostatic attraction and repulsion causes the voltage on the other side to vary correspondingly. This produces an apparent current flow of alternating current through the capacitor. A capacitor blocks direct current while allowing alternating current to pass. The effects of a coil and capacitor in isolating direct current from alternating current components will be further discussed in Chapter 10.

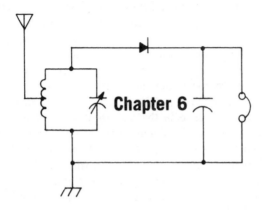

The Basics of Home Wiring

It should be understood right from the start that we do not intend to qualify you to install electric power wiring in your house. This chapter is intended only to show how it is done. Power wiring in your home should only be installed by or under the direct supervision of qualified persons. This is a matter of your own safety and that of all others who might use the newly installed lines. Electrical power in a commercial power system is safe only as long as it is controlled.

To help ensure safety, a national set of standards has been established for home and industrial installations. It is called the National Electric Code, and a copy can be obtained from the National Fire Protection Association, 470 Atlantic Ave., Boston, MA 02210. There is a nominal charge for this publication.

In addition to the National Electric Code, local communities have their own specific regulations which are administered through a Housing, Building, or Wiring Inspector. In some communities, all wiring must be done by a licensed electrician, even in your own home. If you intend to install your own power wiring, touch base with the local inspector and be sure you know all the regulations that apply.

SAFETY GROUNDS

It is common practice to connect one side of the domestic power source to the ground. This is done to protect the generating facilities and intermediate substations from extensive damage that might be caused by lightning. This grounding does, however, have its disadvantages.

Many of the older model power tools and home devices were built with metal enclosures. Any internal short circuit that caused one conductor to touch the metal frame could produce a very dangerous condition. The power plugs were built in such a way that there was no difference in the way they were inserted into the outlets. Thus, either wire might be on the ungrounded side of the source. If the ungrounded side happened to be accidentally shorted to the case of the appliance, and the user, while holding the device, brushed up against a water pipe or other grounded object, the result could be fatal. This writer was sent flying by exactly such an accident in 1965.

To eliminate such dangerous shocks, manufacturers have gone to three-wire power cords which have a safety ground. The third wire is connected to the case of the device, and the plug can only be inserted one way. Any accidental short that connects the ungrounded side of the source to the case will blow the fuse or throw the circuit breaker—a far more desirable condition than killing the user. Many of the older devices with a two-wire power cord are still around, however, so be careful. Be sure that the cord is always in good condition, not cracked or frayed.

Electric power in the home is generally supplied at a potential of 120 volts. High-power devices such as electric dryers, stoves, or home heating systems are supplied 240 volts, to minimize current.

POWER CORDS

As a matter of safety to the user, the outer frame of newer household electric devices is usually made of a non-conductive material, and those that are not have the outer frame electrically connected to the ground. Grounding is accomplished through a rounded, third prong that is rapidly becoming standard on electric appliance plugs. Many older, or simpler, devices still have a two-wire power cord and do not have a safety ground.

Power cords connecting home devices to the domestic source presently use one of the three types of wire. Small devices use rubber or plastic coated wire with the conductors insulated only by the outer jacket, and the plastic-molding process attaches the two coated conductors together so that they can easily be peeled apart. A close inspection will show that the conductors are marked, sometimes by one being made of light-colored metal, one having a tracer

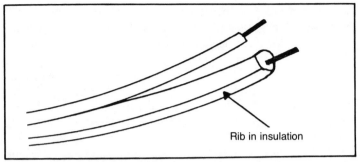

Fig. 6-1. Two-wire lamp cord. Note the rib, or ridge in the insulation to help keep track of which wire is which.

(ridge) marked on the outer insulation, or by a colored thread among the wires of the conductor, see Fig. 6-1.

The power cord for larger appliances or power tools is much more rugged in construction than the simple "lamp cord" just described. The two conductors are rubber or plastic coated, and are not attached to each other. They are twisted together and then wrapped with a paper strip. The whole wrapped assembly is then coated with a plastic or rubber outer covering. The two conductors are differentiated from one another by the color of the inner insulation. Usually, one is black and the other white, see Fig. 6-2.

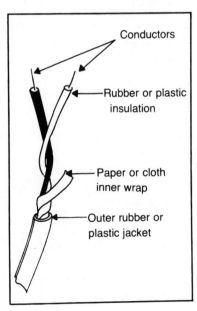

Fig. 6-2. Construction of general-purpose power cord.

Cord of this type may contain two or three conductors. The third conductor is usually a safety ground, and should be green.

Devices that are a heating element are supplied with a special wire that has fireproof insulation. Prior to the development of glass-fiber, asbestos was used in this application. The two power-carrying conductors were insulated with the fireproof material, and the safety ground often had conventional insulation. The outer wrap consisted of a woven fabric, which was not necessarily fireproof, Fig. 6-3.

Power plugs are often molded onto most appliances, although heating devices using fireproof cord are impractical to attach the plug in that way. Simpler devices use the familiar two-prong plug. Others use a plug with a third, rounded prong which goes to the safety ground. When a plug of this type is to be inserted into one of the old, two-wire outlets, an adaptor allows connection of the safety ground. *There is no protection when using such an adaptor unless the ground wire is connected.* See Fig. 6-4.

BASIC CIRCUITS

Except for electronic devices (radios, TV sets, stereos, etc.), practically all household electric devices have a very simple circuit. The power source is connected through a switch to the load. The load is usually a motor, light bulb, or heater, see Fig. 6-5. The frame of the device is connected to the safety ground when it is used.

The simplest of all household devices is an electric lamp. The switch is usually built into the lamp fixture. All one has to do is

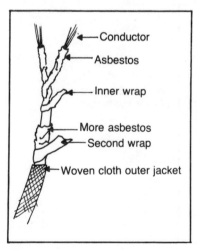

Conductor
Asbestos
Inner wrap
More asbestos
Second wrap
Woven cloth outer jacket

Fig. 6-3. Asbestos-insulated cord for heating appliances. This type of cable is no longer used in new equipment.

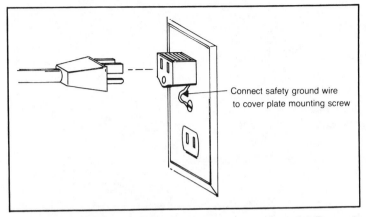

Fig. 6-4. Use of an adaptor to plug 3-wire cord into a two-wire outlet. Be sure to ground the safety ground wire.

connect the wire for the power source to the two terminals on the fixture, Fig. 6-6.

Since the outer case of a lamp is seldom made of metal or conductive material, a two-wire power cord can be used. If the lamp has a metal frame, a three-wire cord is used and the safety ground (green wire) is connected as shown in Fig. 6-7.

A household power system in the early years consisted simply of two wires entering the house from the poles. One of the wires was connected to the ground, the other was the "hot" lead. These were brought in, usually to the basement, through an iron pipe called a conduit. A kilowatt-hour meter kept track of the amount of power used, and a system of fuses protected the various branches which distributed power throughout the house, see Fig. 6-8.

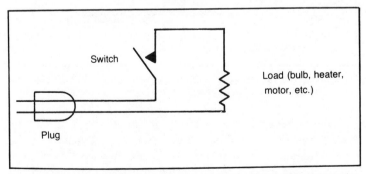

Fig. 6-5. A schematic diagram of almost every basic household electric device. Not much to it, is there?

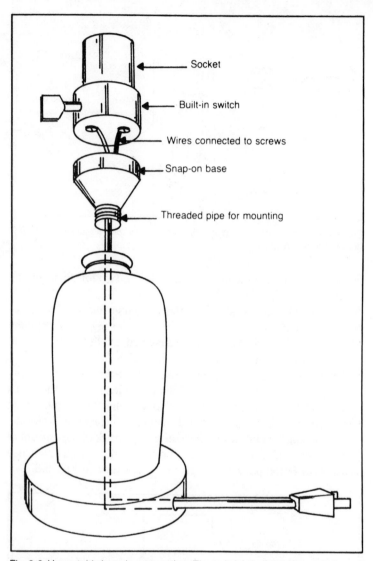

Fig. 6-6. How a table lamp is put together. The switch is built into the socket, so all you need worry about is connecting the two wires.

Wiring techniques were crude, to say the least, in the very earliest systems. The wire itself was not concealed, but ran in plain sight straight across walls and ceilings. It wasn't very decorative—heavy, black wires supported away from the wall by white, porcelain insulators.

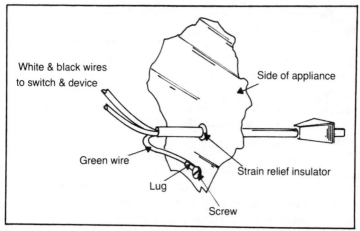

Fig. 6-7. A safety ground is usually connected to the metal framework of an appliance just where the cord comes in.

There are those who feel that the old, open wire technique was the safest ever. However, because the wires were all out in the open, they were far more vulnerable both to damage and to tampering by untrained persons.

As the industry progressed, BX cable came into use. BX was the name of a common flexible, steel-jacketed cable that became the standard for home wiring just before the second World War.

It was miserable stuff to work with. You had to first use a hacksaw to cut the spiral steel jacket, peel back the steel spiral (which often caused many a cut finger), cut off the excess jacket, then cut the wires themselves.

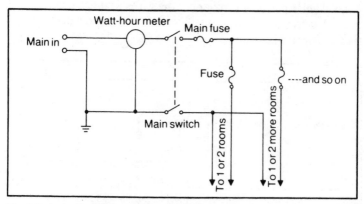

Fig. 6-8. A simple, old-time power distribution system.

Each outlet (light or plug connection) was connected to the wiring system through a metal box. The outlets or junction boxes were provided with holes that could easily be punched out, and clamps were available to secure the metal jack of the cable to the metal box. Thus, the entire electrical wiring of the home was surrounded by a grounded, steel armor, and there wasn't very much that could happen in the way of accidental damage. Although different cable is often used today, the metal junction and outlet boxes, with a variety of input clamps, are still the standard way.

A typical household service now uses a 230-volt line coming into the house. This comes in on three wires, the grounded wire being connected to the center tap of the power company's transformer, mounted on the nearby pole. You therefore have two 115-volt lines coming in with a single, common line for their return. The return is connected to ground, see Fig. 6-9.

Inside the house, this kind of service provides a 230-volt line to feed electric stoves, major appliances, etc. Then, either side of the #230-volt line can be used as a 115-volt line with respect to the common wire, thereby providing dual 115-volt service for distribution throughout the house. The various rooms in the house are so arranged as to distribute the load as evenly as possible between the two halves of the service, Fig. 6-10.

WIRING TYPES

Three types of wiring schemes are used today. One is the armored cable already discussed. The second is a type of cable generally called roamex. The third is individual conductors.

Roamex normally comes in two types. One type has the wires embedded in plastic, the other has the wires wrapped together with a tough fabric. The first is suitable to bury in the ground; the second is for indoor wiring.

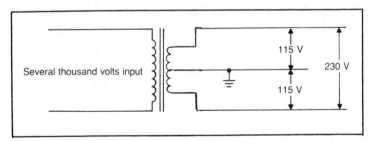

Fig. 6-9. How a center-tapped transformer can provide a 230-volt line and/or two 115-volt branches.

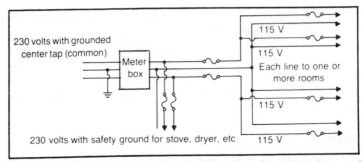

Fig. 6-10. A simple home power distribution system. Sometimes the actual connection to ground is made as shown, and sometimes it's made from the other side of the meter. It all depends on local wiring codes. When in doubt, ask the town wiring inspector.

Either roamex or armored cable is available in a variety of conductor sizes, and as either two- or three-wire line.

Three-wire line can be made with the third conductor bare, or with the third conductor insulated, but slightly smaller than the other two. When the third conductor is insulated, it is usually colored green, representing the ground (safety) wire. The white conductor also connects to ground, but is still the return wire for the circuit and carries current.

A third scheme of wiring uses individual insulated conductors running through a metal conduit. This is most frequently used in industrial installations, although you do occasionally see it in a home, particularly in the newer ones. It is the most expensive, but also the safest of the three.

BOXES AND FIXTURES

Wherever wires must be connected together, or wherever an outlet or light fixture is to be plugged or wired into the system, a box (usually metal, but sometimes plastic) is used to contain the connection. This is true no matter which type of wire is used. The boxes themselves are, more often than not, made to be adaptable to any kind of wiring. They are made with punch-out holes, into which standard fittings can be set to accept the wire used. In all cases, the fittings are secured to the box and to the wire or conduit. This is to prevent any accidental strain from being applied to the connections inside.

Fixtures for outlets or switches all have a standard mounting scheme using two screws that mate with lugs on a metal box. Since the mounting is standard throughout the industry, different brands are interchangeable. Boxes that house only wire interconnections are covered by a blank metal plate.

An outlet fixture (such as a socket) has one brass and one steel connection. The brass screws are for the hot wire (usually black), and the steel screws are for the grounded, white (also called "common") wire. Three-wire-outlet fixtures have a third connector painted green, which is for the safety ground.

FUSES AND LOADS

In the following description, the term "fused" will be used in reference to protection by either a fuse or by a breaker. The 230-volt service is brought into the house through the power meter to the input panel where a fuse or breaker box is mounted. Connection to ground is usually made at this point. The service is then divided into two 230-volt branches. One serves the electric stove or dryer, the other feeds the rest of the house.

In the second of the two 230-volt branches, the center-tap feature of the service is wired to provide two 115-volt lines which supply the rest of the house. Effort is usually made to balance the average load equally between the two branches.

Both 230-volt branches are fused independently of each other. Each of the branches of the two 115-volt lines is also fused, see Fig. 6-10.

In calculating the load for any given line, one must take into account the power demand of each of the various devices to be used. Table 6-1 gives the average demand of each of the better-known household appliances. The total of all the power demands on any given line must not exceed the amount for which the line is fused.

When using this table, you can simply add the current demand of the devices in question. If, for example, the line is fused for 15 amps, the total of all the current demands on that line must not exceed 15 amps. If it does, and all the devices happen to be on at the same time, the fuse will blow.

Fuses or breakers are made for one reason only, and that reason is the protection of the user. Those who defeat the purpose of a fuse or breaker are inviting trouble. If the fuse blows frequently, the line is overloaded. If the fuse is bypassed to get the lights back on, the wiring can overheat, and that can start a fire. Play it safe and call an electrician.

Any discussion of overloaded electric wiring would be very incomplete without mention of two prime villains: cube taps and

Table 6-1. Power & Current Demand of Common Electrical Devices.

Range	⎫	12,000 - 16,000 Watts	64 - 72 amps
Water Heater	⎬ 220 V	2,000 - 3,500 Watts	9 - 15 amps
Dryer	⎭	8,400 - 9,000 Watts	40 - 41 amps
Air Conditioning		600 - 4,700 Watts	5 amps @ 115 V to 20 amps @ 220 V
Room Heater		1,800 - 7,000 Watts	15 - 60 amps
Microwave Oven		1,500 - 2,000 Watts	13 - 17 amps
TV		300 - 500 Watts	2.5 - 4 amps
Washer		500 - 800 Watts	4 - 7 amps
Toaster		approximately 1,000 W.	about 8.7 amps
Iron		approximately 750 W.	about 6.5 amps
Ceiling light		approximately 100 W.	about 0.3 amps
Table lamp		40 to 60 Watts	0.3 - 0.5 amps
Stereo		100 - 200 Watts	0.8 - 1.7 amps

extension cords. Either can be quite safe if used with care; either can be disasterous if it is misused.

A cube tap (also called a three-way outlet) enables several appliances to be plugged into the same outlet. It should only be used when it is absolutely necessary, and should never be part of a permanent setup. When a cube tap is used, care should be taken to see that the total current of all the appliances plugged into it does not exceed the rated current of the outlet in which it is used, and that the overall circuit is not overloaded.

An extension cord allows the device to be farther from the outlet than the length of the standard line cord. That, in itself, is not so bad. The trouble can lie in two possible faults. First, most household extension cords have a multiple socket at the output end. This can lead to the same overload problems as a cube tap. Next, since they are very small and portable, extension cords tend to take more of a beating than appliance power cords or permanent electric outlets. The contact efficiency degrades very easily, resulting in increased resistance. When current flows through a resistance, heat is generated. Moreover, loss of voltage through the contact resistance of a faulty extension cord could, in time, damage the appliance. The safest rule is, if the connections to the extension cord get warm, don't use it.

SERVICE

Servicing electrical equipment requires experience and knowhow. However, a few of the simpler repairs can be done with a minimum of experience and, as long as one is careful, they can be done by a nonprofessional person.

It is a prime rule that the power *must* be turned off before attempting any kind of repair, even to the power cord of a simple appliance.

All connections must be mechanically tight, particularly those that carry large amounts of current. A heavy current can produce heat in a relatively small resistance. Furthermore, the wire itself must be clean. A little oxide can result in enough contact resistance to cause heat. Clean the wire and the screw terminals by scraping with a knife or by wiping with fine emery paper until the metal parts are shining.

Replacing a Power Plug

This is the simplest of repairs, one that every housewife should know. All it takes is a replacement plug, available at any hardware store, and (depending on the kind of plug), a screwdriver.

There are two kinds of plugs in common use today. One is made to use with double lamp cord, and has prongs that dig through the insulation to make the contact. All that is needed with this kind of plug is to trim the ends of the cord even, open up the plug, and insert the cord. Then, when the plug is reassembled, it's ready for use. See Fig. 6-11.

Larger appliances that use black, rubber-jacketed cable use a replacement plug that has screw terminals. Older appliances use a two-wire cord, while newer ones use three-wire cable. Except for the presence of a third terminal, the wiring procedure is roughly the same for either cable, see Fig. 6-12.

Note from the illustration that the wires go around the prong before being connected to the screw. Strain relief is provided by knotting the wires in an underwriter's knot, Fig. 6-13. You must first strip about ½-inch of insulation from the ends of each wire, twist the strands together, and wrap the wires around the respective screws. Then tighten the screws, insert the insulator, and the

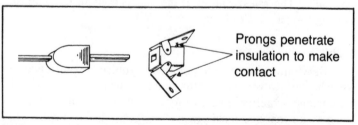

Prongs penetrate insulation to make contact

Fig. 6-11. A do-it-yourself's delight. A simple, snap-on plug. You don't even have to strip off the insulation. Two sharp prongs penetrate the insulation and make the contact.

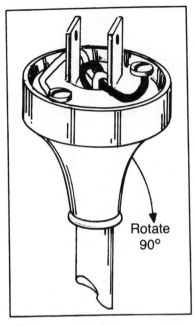

Fig. 6-12. A heavy-duty, replacement plug. Note how the wires wrap around the terminals for strength, and the "underwriter's" knot.

Rotate 90°

job's done. Note that the wire goes around the screws in the same direction the screws will be turned when tightened (clockwise).

Three-wire plugs do not use the underwriter's knot for strain relief. Many have a clamp or other means of strain relief. Many have a clamp or other means of strain relief built in. With this kind of plug, however, you must watch the color of the wires. The white wire goes to the "silver" terminal, the black one goes to the brass

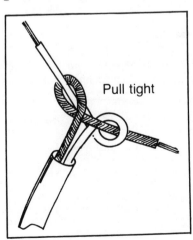

Pull tight

Fig. 6-13. How to tie an underwriter's knot.

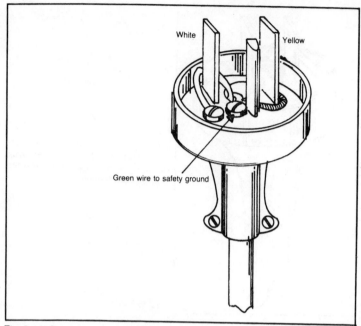

White   Yellow

Green wire to safety ground

Fig. 6-14. Connections to a three-wire plug. The green wire goes to the rounded, ground post. The white wire goes to the silver terminal. The black wire goes to the brass terminal. Don't connect it any other way.

terminal, and the green wire goes to the rounded, protective-ground terminal. Otherwise, the technique is similar. Route the wire around the prong to the screw. Be sure the correct length is stripped, and that the strands of wire are twisted together. See Fig. 6-14.

Lamps

Most incandescent-light, wall-lamp, and table-lamp units have a fixture with the switch built in. Those that don't either have a

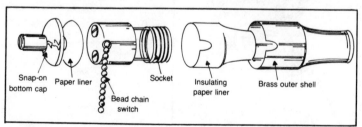

Snap-on bottom cap   Paper liner   Bead chain switch   Socket   Insulating paper liner   Brass outer shell

Fig. 6-15. Exploded view of a basic light fixture.

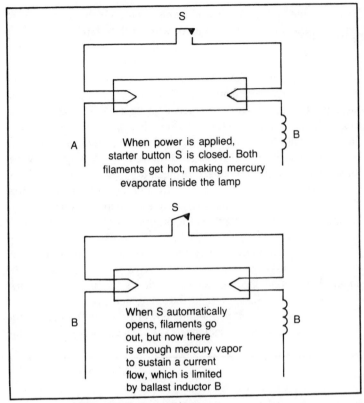

Fig. 6-16. Operation of a fluorescent lamp.

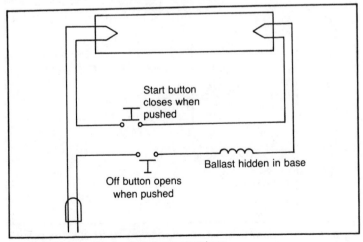

Fig. 6-17. Schematic of a desk fluorescent lamp.

switch somewhere on the wall or elsewhere on the assembly. At any rate, the socket, either with or without the switch, is relatively easy to replace, see Fig. 6-15. Just remember to be sure the power is *off*.

The light unit might be built in a variety of schemes in which the socket is secured in any one of a number of ways, none of them overly complicated. Once it is mechanically loosened, the electrical connections to the socket are obvious. In a wall or ceiling fixture, be sure the black wire connects to the brass terminal, and the white wire to the white-metal terminal. If there is a safety ground wire, there will be a green-colored terminal for that.

Fluorescent lights are a different circuit entirely from incandescent-light circuits. Two circuits come into play for this kind of light, one for the starter and the other to sustain the light after it has started up.

In the startup circuit, two filaments are connected in series, one at each end of the lamp. These serve to heat the gas inside the lamp. After a few seconds, an automatic switch in the starter unit opens to put the second circuit into play, see Fig. 6-16.

The second circuit applies the power between the ends of the lamp through a series inductance. The inductance both limits the

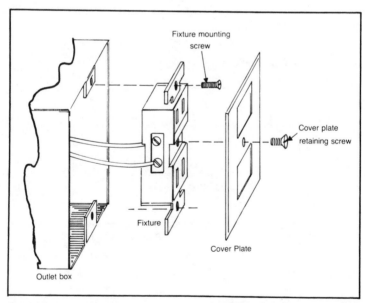

Fig. 6-18. Electric outlet assembly. BE SURE YOU TURN OFF THE POWER BEFORE YOU OPEN UP ANY OUTLET OR LIGHT FIXTURE.

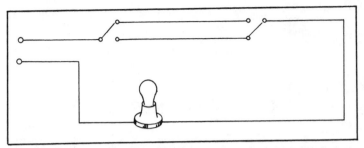

Fig. 6-19. How two single-pole double-throw switches can control a light from two places. Trace the circuit through. If either switch is changed from the positions shown, the light will turn on. Then, if either switch is thrown, the light will turn off.

current through the gas inside the lamp, and provides a "kick" to start the current flow.

The series inductance inside the lamp is called the *ballast*. It is the ballast that is responsible for the humming sound sometimes heard in fluorescent fixtures. As long as this humming is not so loud as to be objectionable, there is no need to replace the ballast.

If the lamp glows and flashes, or if the ends are visibly blackened, the bulb should be replaced. *Use extreme caution when handling fluorescent bulbs.* They are coated inside with a very poisonous phosphor. If you are cut by a piece of a broken fluorescent lamp, no matter how slightly, have a doctor check it out.

If even the slightest glow in the lamp end is visible when the lamp is turned on, chances are that the starter is all right. Most of the problems with fluorescent lamps involve either the starter or the bulb. The ballast seldom goes bad, but when it does you'll know by the odor!

In fluorescent desk lamps, and some wall lamps, the starter function is accomplished manually by a pushbutton. These lamps have a separate "on" and "off" button. See Fig. 6-17.

Wall-mounted switches and electric outlets can easily be replaced by turning off the power and then removing the cover plate. The switch or outlet unit is then removed by undoing two mounting screws, after which the terminals are readily accessible. Make a note of which wire connects to which terminal, and connect the new fixture the same way, see Fig. 6-18.

Lights that are controlled from two independent switches use double throw switches connected as shown in Fig. 6-19.

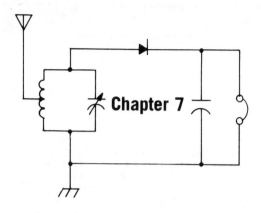

Electricity and Sound

This chapter covers the behavior of electrical devices that either capture or produce sound. In order to better understand such devices, it is necessary that the nature of sound itself be understood. Sound is the result of an object vibrating and causing a displacement of its environment. That environment can be air, water, or any such substance, so long as it has some amount of elasticity. The speed at which sound travels through a substance is proportional to its elasticity. Table 7-1 illustrates this principle.

Sound waves result from a vibrating object compressing and stretching the substance, which we will hereafter refer to as the medium. These waves travel through the medium and, when they reach another object, cause that object to vibrate, see Fig. 7-1.

The wave length depends on the speed with which sound travels through the medium, and on the frequency of the vibration. It is the frequency of the vibration that determines the pitch of the sounds we hear. Table 7-2 shows the relationship between pitch and frequency of the notes on the musical scale:

The human ear is capable of detecting sounds at frequencies ranging from about 20 cycles per second to about 16,000 cycles per second. The exact limits of that range varies from one person to another. The ear has its greatest sensitivity at about 2300 cycles per second, which is about an octave over the E above high C on the musical scale.

Table 7-1. Speed of Sound Through Different Materials.

Approximate Velocity of Sound in Various Substances	
Medium	**Velocity**
Air	1100 ft/sec
Carbon dioxide	846
Water	10900
Pine	4708
Maple	13470
Oak	12620
Rubber	177
Granite	19685
Iron	17390

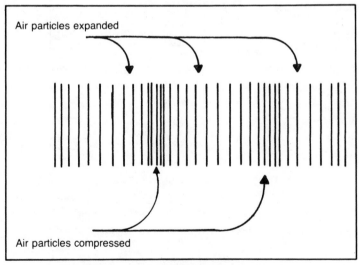

Air particles expanded

Air particles compressed

Fig. 7-1. This exaggerated drawing demonstrates how sound waves are made of air particles alternately crowding together and pulling apart.

Table 7-2. The Musical Scale.

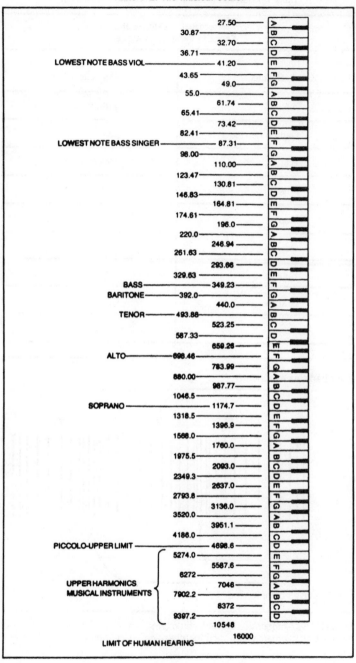

So far, we have discussed sound in the terms of pure tones. Outside the laboratories, pure tones are seldom heard. The sounds we hear are often mixtures of several tones, or they may consist of a pure tone distorted by any number of harmonics.

A harmonic of a given frequency is any exact multiple of that frequency. For example, if the fundamental tone is 1000 cycles per second, its second harmonic is 2000 cycles per second; its third harmonic is 3000 cycles per second, and so on.

The number of harmonics, and their intensity relative to the fundamental frequency, gives the individual quality to a sound source. An organ can produce a more nearly pure tone than any other instrument, while a trumpet or a saxophone playing the same note as the organ will make a sound rich in harmonics.

If sound waves, travelling through one medium, such as air, reach another medium of very different elasticity, such as a wall, part of the energy will set up waves in the new medium while the remainder will be reflected (echo). In a closed room, the sound will rebound from wall to wall until the energy dissipates. This action is called reverberation, and governs the sound characteristics of the room.

The speed with which sound travels through air depends on several factors, particularly the temperature. At 70° F, (20° C) the speed is 1100 feet per second.

THE STORY OF THE TELEPHONE

Right after the telephone had been invented, an envious scientist stated, "If Mr. Bell had known anything at all about electricity, he would never have been able to invent the telephone!"

The man's outburst was really more correct than you may think. Bell was a teacher of the deaf, and his greatest knowledge was in the nature of sound and hearing.

Bell knew that, in order to faithfully reproduce sound with electricity, he must find a way to cause rapid, minute variations of current following exactly the rapid, minute variations of pressure which our ears interpret as sound. His efforts were directed toward creating this "undulating current" as he called it.

If the current could once be made to undulate, it would simply be necessary to place a metal diaphragm where it would be influenced by an electromagnet, Fig. 7-2.

The undulating current in the electromagnet would place a varying strain on the diaphragm, and the resulting motion would produce sound. There's no need to tell you it worked. In fact, it

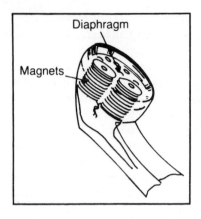

Diaphragm

Magnets

Fig. 7-2. Sound is produced in a telephone by electromagnets pulling on a thin metal diaphragm.

worked so well that the same principle is still used on most telephone receivers and, with slight modification, is the principle associated with loudspeakers.

The hard part of Bell's work was the task of capturing sound waves and causing them to produce an undulating current. After much hard work and many experiments, he hit on the idea of a wire immersed in a sulfuric acid solution. As the wire, which was connected to a diaphragm, vibrated in and out of the acid solution, the resistance of the device would vary, resulting in a varying current. See Fig. 7-3.

He built his prototype and connected it to a reproducer (receiver). While his assistant was carrying the receiver into the other room, Bell reached for some acid. He slipped and spilled the acid onto his hand. In pain, he called out, "Mr. Watson, come here. I want you!"

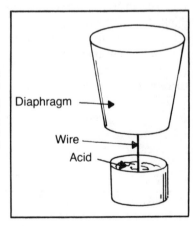

Fig. 7-3. Bell first imposed voice onto an electric current by attaching a wire to the center of a diaphragm. The wire hung into a container of acid. Vibration of the diaphragm moved the wire up and down in the acid, making a varying resistance which, in turn, varied the current.

Diaphragm

Wire

Acid

The assistant came on the run. After a few breathless words, Bell's pain was forgotten. Mr. Watson had heard the voice from the apparatus, not directly through the air. Bell's invention was a success, and the very first message it had transmitted was an emergency call for help. A complete telephone circuit is shown in Fig. 7-4.

The experimenter who wants to duplicate Bell's accomplishment need not fool around with sulfuric acid. There is a better way, based on the principle used in most telephones today, as well as in a great many communications microphones. The microphone is called a carbon-button type, shown in Fig. 7-5.

A diaphragm, which picks up the sound vibrations, is one side of a small container filled with carbon granules. As the diaphragm vibrates, it alternately packs the granules more closely and more loosely. This results in an undulating current just as did the rod immersed in sulfuric acid. What's more, it's much safer.

TELEPHONE REPRODUCER

This project is not likely to be what one might call a thing of beauty unless you're good at hand fabrication. However, it does work. It may not be too efficient, but it works. Materials needed:

About 5 square inches of steel shim stock, 2 to 3 mils thick.
1 pound No. 30 to No. 36 enameled copper wire

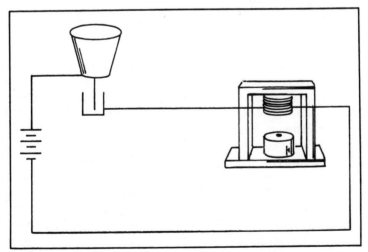

Fig. 7-4. Bell's circuit was quite simple. His transmitter varied the current in an electromagnet, which pulled on a lever to vibrate a diaphragm.

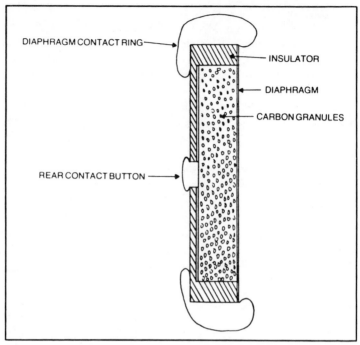

DIAPHRAGM CONTACT RING

INSULATOR

DIAPHRAGM

CARBON GRANULES

REAR CONTACT BUTTON

Fig. 7-5. Construction of carbon microphone button.

Empty typewriter-ribbon spool
Small metal can
Record player

Wind the wire onto the typewriter ribbon spool. The more you can get onto the spool, the better. Leave about 6 inches at each end for connections. This will be the magnet for the reproducer.

Mount the spool in the center of the can. Position it carefully so that the top surface of the spool is about 1/64 of an inch below the top edge of the can. Drill two small holes in the bottom of the can and bring the wires out. Use some small plastic tubing to protect the wire.

Cut the shim stock into a disk the same diameter as the rim of the can. Carefully place the disk over the top of the can, and secure it into place with a few drops of contact cement. The shim stock should be as close as possible to the spool without actually touching it. Assembly details are shown in Fig. 7-6.

Find an old record player, preferably the kind that has tubes rather than transistors. Disconnect the speaker wires. Run a pair of

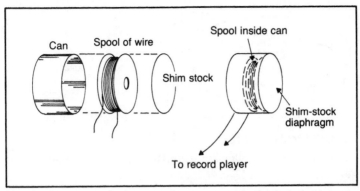

Fig. 7-6. It won't be very loud, but if you build it carefully, it'll work. Mount the magnet coil in the can so that it comes very close to the diaphragm, but does not actually touch it.

wires from the speaker connections of the record player to the leads of the sound reproducer, see Fig. 7-7.

Put on a record and turn up the volume. If you have done everything right, you should hear sound from the reproducer.

The reproducer you have built is very similar to the earphones still used in some of the less expensive headsets, and in your telephone. Electric impulses from the record player magnetize the coil and attract the shim stock (diaphragm) causing it to vibrate. The vibrating diaphragm produces sound. If yours has turned out to be very efficient, I suggest that you save it for the next experiment.

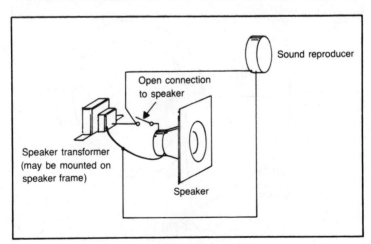

Fig. 7-7. How to connect your home-brew sound reproducer to an old record player.

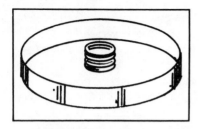

Fig. 7-8. Hose washers mounted inside the metal jar cover.

A CARBON MICROPHONE

Gather up the following materials to make a simple microphone:

Metal lid from peanut-butter jar
2 or 3 mil brass shim stock, same size as jar lid.
3 garden-hose washers
Felt ring, same size as the hose washers
Old flashlight batteries (2)

Scour the inside of the jar lid until clean, bright metal is exposed. Mount the hose washers in the center as shown in Fig. 7-8.

You may have to make the felt ring yourself. It should be about the diameter of the washers, and the center opening should be about the same size as the washers. Glue the felt ring on top of the

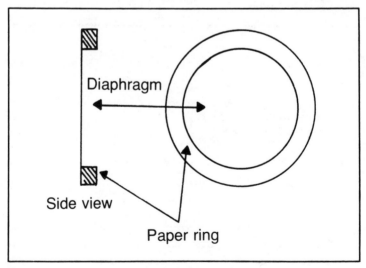

Fig. 7-9. Paper ring glued to the shim-stock diaphragm.

washers. When the assembly is complete, it should be no higher than the rim of the lid.

The diameter of the shim stock, which will be the diaphragm, should be slightly larger than the rim of the lid. Solder a short length of wire to the edge of the shim stock, and another to the jar lid.

Cut a ring of paper the diameter of the diaphragm. The inner opening should be only slightly less than the inside of the jar lid, see Fig. 7-9. Glue the paper onto the diaphragm, and give it a light coating of varnish.

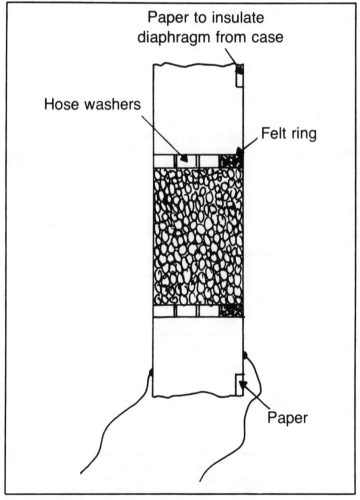

Fig. 7-10. Side view of your home made carbon microphone.

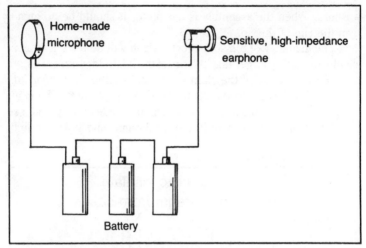

Fig. 7-11. Whether you use home-brew or manufactured parts, this is the way to connect a simple telephone system.

Open the old flashlight cells to recover the carbon rods. You will have to be careful with this, as the carbon rods break easily. Use a file to reduce the rods to powder. You will need enough to fill the inside of the hose washers about four fifths full.

After the carbon powder has been put into the washers, coat the top of the felt ring and the rim of the lid with contact cement. Very carefully press the diaphragm into place as shown in Fig. 7-10.

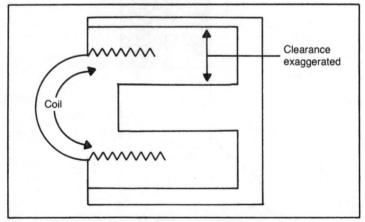

Fig. 7-12. A dynamic microphone has a diaphragm with a coil attached, suspended within the field of a specially-shaped magnet. The vibrating diaphragm moves the coil within the magnetic field, thus generating electricity.

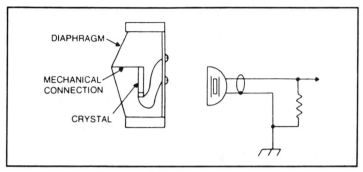

Fig. 7-13. Crystal microphone construction & circuit.

Just how well the microphone works will depend on how well it has been built. Connect it to a sensitive earphone with a telephone cell or other battery to power it and try it out. Connections are shown in Fig. 7-11.

What you have just done is build a telephone system, in its simplest form, from the bottom up. If you had two microphones and two earphones, there is no reason why they couldn't be mounted into a single telephone handset. That, plus a fiendishly intricate switching and amplifying system, is basically what Ma Bell is still using today. Of course, the carbon button is not the only kind of microphone. There are a great many different kinds. One of the most popular is the dynamic. A dynamic microphone has a very lightweight coil of wire fastened to the center of the diaphragm. The coil fits over the poles of a specially shaped magnet, see Fig. 7-12. As the diaphragm vibrates, the coil moves between the magnet poles. This generates tiny amounts of electricity which are then amplified.

One of the more interesting microphones employs a thin piece cut from a mineral crystal. This crystal has a characteristic in which,

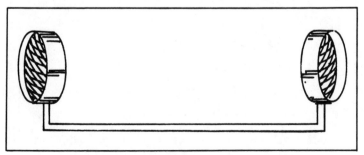

Fig. 7-14. Interior of a crystal phono pickup.

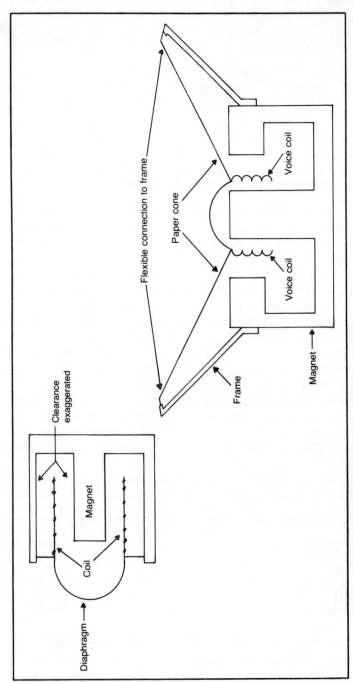

Fig. 7-15. Comparison between a dynamic microphone and a loud speaker.

if mechanical pressure is applied, a voltage is produced. The diaphragm of a crystal microphone is mechanically connected to the crystal (Fig. 7-13), and the vibration of the diaphragm causes voltages to be generated. This phenomenon is called piezoelectricity.

SOUND-POWERED TELEPHONE

Here's a simple experiment. You'll need two inexpensive crystal microphones and about 50 feet of lamp cord.

Wire the two microphones together as shown in Fig. 7-14. You will find that either one can act as a microphone or as an earphone. Since they produce their own electricity, batteries are not needed. This experiment can also be done with dynamic microphones. However, the output level may be very low. This is because, except for the more expensive models, a dynamic microphone has a very small diaphragm.

A dynamic microphone is very similar both in construction and in principle to a loudspeaker. Most loudspeaker diaphragms are paper cones, and have much greater area. In both instances, we have a lightweight coil of wire, attached to the diaphragm, suspended between the poles of the magnet see Fig. 7-15. Not only will a loudspeaker function as a microphone in a pinch, but it is often used as both in intercom systems.

These are just a few of the many kinds of microphones in use today. Except for the carbon microphone and the capacitor microphone, which are in classes by themselves, most microphones operate either on an electromagnetic or a piezoelectric principle.

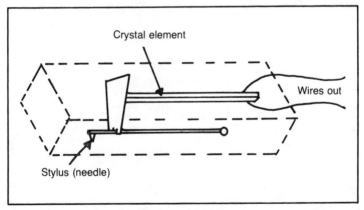

Fig. 7-16. Two crystal microphones can be connected as a sound-powered telephone system. All you have to do is wire them together. What could be simpler?

PHONOGRAPHS

Next, on the subject of the relationships between sound and electricity, we should discuss phonograph pickups. A phono pickup is, after all, very much like a microphone. The main difference is that it has no diaphragm. The moving element is attached instead to a needle. As the needle rides in the record grooves, irregularities in the sides of the groove cause the needle to vibrate. Except for the voltage levels coming out, the amplifier would never know the difference between a microphone and a phono pickup of the same type. Figure 7-16 shows a crystal phono pickup.

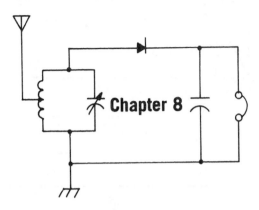

The Electronic World

So far, this book has covered a few of the various uses for direct current and alternating current. One advantage of the latter has been the ease with which voltage levels can be changed without loss of power. How convenient it would be if there were some way to convert alternating to direct current and back again.

There is. Direct current, as we have seen, can be interrupted by a vibrator (buzzer) and fed into the primary of a transformer. When this is done, alternating current comes out of the secondary. Now the problem is to convert the other way. That is, to go from alternating to direct current.

Let us review the difference between the two. Direct current flows in one direction only. While the voltage level may vary, the direction of the current flow remains constant. Alternating current, on the other hand, not only has the voltage level vary, but also changes direction, current flowing first one way then the other.

When we turn on a direct-current circuit, the voltage rises to the maximum level and stays there or, depending on what the equipment does, may vary in level all the way from zero to maximum, but never reversing direction. When we turn on an alternating-current circuit, the voltage rises from zero to maximum, then falls, not just to zero, but below, assuming a negative value which reverses the direction of current flow, see Fig. 8-1. This constant reversal of direction is of little consequence in

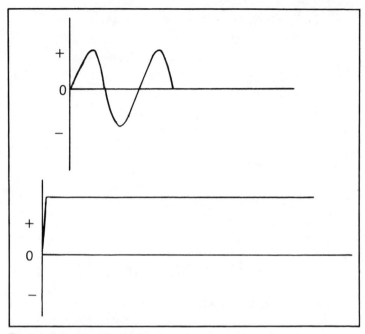

Fig. 8-1. Alternating *vs* direct current.

most applications, but there are situations where it can be of a distinct advantage.

With alternating current, we can, by using a transformer, set the voltage to the approximate level we need. Once the voltage level is established, we need some way to ensure that the direction of current flow will be constant.

EARLY DIODE EXPERIMENTS

The device that accomplishes this is called a rectifier or diode. It was this seemingly trivial thing that opened the gates to the world of electrical and electronic technology.

The first rectifying device was discovered by accident. Shortly after the invention of the first incandescent lamp, Thomas Edison was attempting to find ways to prevent the gradual accumulation of a dark film on the inside of the bulb. (This still occurs in present-day lamps.) In the course of his experiments, Edison tried partially enclosing the lamp filament in a metal shield. This shield was supported by a wire that protruded through to the outside of the bulb. Edison noticed, quite by accident, that a small voltage existed between the shield and the filament, see Fig. 8-2.

At this point, we are sad to note, the wizard of Menlo Park apparently blew it. He noted his discovery, but continued on in his search for a cure to the original problem and never followed up on this new discovery.

Edison's discovery eventually came to the attention of an English scientist named Lee Fleming. Fleming, who was working for Marconi, the inventor of radio communication, applied the principle to a device for detecting radio waves—which was simply rectification of very weak high-frequency alternating current. It was called a valve because it could turn current flow off and on.

While Fleming's "valve," later to be called the vacuum-tube rectifier, remained the most widely used rectifying device for a great many years, the device that eventually replaced it, or at least the predecessor of that device, was invented almost immediately afterward. American interests in radio communication preferred having their own rectifying device, rather than pay royalties to Marconi's company. Enter yankee ingenuity. Within a few years, two other devices for rectifying alternating currents were on the market.

One of these was a chemical device called the electrolytic rectifier. It consisted of a container of liquid in which two electrodes were immersed. One of the electrodes was a fine platinum wire resting on the surface of an acid (electrolyte), and when the applied voltage was of the correct polarity, insulating bubbles would form

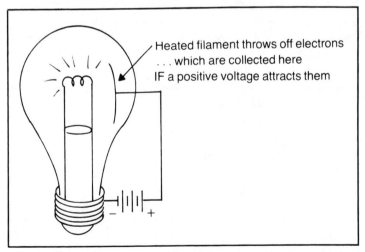

Fig. 8-2. Early in the history of electric lighting, Edison discovered that current could flow through a metal plate within a light bulb close to the filament.

on the surface of the electrode. When the polarity reversed, the bubbles would burst. The other electrode, being of a different metal, would not form the bubbles. The result of all this was that current could flow in one direction, but not the other, see Figs. 8-3 and 8-4.

The electrolytic rectifier had several disadvantages. The back resistance would allow some reverse current to flow, making it very inefficient. Also, there was always the danger of a spill, and the electrolyte could cause a considerable amount of damage.

While the vacuum diode predominated for a long time, the device that eventually took over was invented while the art was still very young. Although it was useless for any appreciable amount of power in its early years, the solid-state crystal diode quickly took over as a radio detector and held the limelight until we learned how to make Fleming's valve into an amplifying device.

The diode featured a small piece of lead sulfide (galena) with a thin wire (catwhisker) lightly contacting the surface. This was the heart of the crystal set that Grandpa likes to tell you about. I'll tell you how to build one a little later in the book, but first let's get a bit more familiar with the use of diodes as power rectifiers.

VACUUM DIODE

We have previously used the word "diode" in reference to rectifying devices. The word, however, didn't come into use until the invention of vacuum tubes. It means literally, "two elements," and refers to the fact that the original device consisted only of a filament, also called the *cathode*, and a plate, also called the *anode*.

Here's how the vacuum diode worked: When the filament of a lamp is heated, it throws off electrons into the surrounding space.

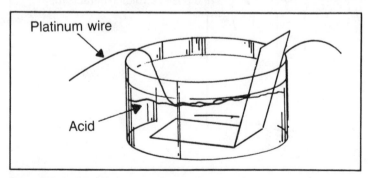

Platinum wire

Acid

Fig. 8-3. One of the earliest radio detectors consisted of a thin, platinum wire laid on the top of a dish of acid.

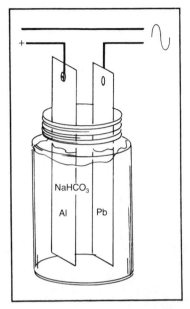

Fig. 8-4. For heavy-duty applications, aluminum and lead electrodes were immersed in a jar of baking-soda solution.

Now, an electron represents a negative charge. If a nearby electrode is given a positive charge with respect to the filament, that electrode will attract the negatively-charged electrons. Consequently, current will flow through the device. If the electrode is given a negative charge, it will repel the electrons, and no current will flow.

Now, suppose we take one of these devices and connect it in series with an alternating-current circuit. On that half of the cycle in which the electrode called the anode is positive, the device conducts; when the anode is negative, it does not conduct. The result is that a series of pulses, all going in the same direction flows through the circuit. As this device developed, another electrode called the cathode was developed. This consisted of a thin metal sleeve wrapped around the filament. The filament heated the cathode, which in turn emitted the electrons. Current flowed only during that part of the cycle in which the plate (anode) was positive. A rectifier and a drawing of its output are shown in Figs. 8-5 and 8-6.

At this point, note that the current flowing through the circuit consists of pulses of current. Most of the time, we need a steady flow of current. Some means must be devised to smooth out those pulses when the application demands it.

The vacuum rectifier worked. It was, for many years, the predominent means of conversion from alternating to direct cur-

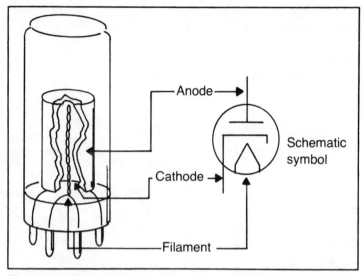

Fig. 8-5. A vacuum-tube rectifier, the modern version of Fleming's invention. Although the early tubes didn't have it, we show the cathode in this illustration.

rent. In fact, there are many old-time radios available on the second hand market that use it. It did, however, have its disadvantages.

One disadvantage was that the cathode-to-anode resistance was very high, resulting in very poor efficiency where low-voltage/high-current applications were concerned. The other was that an external power source was needed to light the filament, thereby warming the cathode so that it could emit electrons.

In addition to all this, the vacuum rectifier needed a relatively high voltage applied to the anode in order to operate. Low-voltage

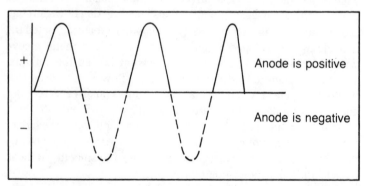

Fig. 8-6. Be it vacuum tube or some other device, a single rectifier cuts off half the ac wave. That's why it's called a half-wave rectifier.

power supplies and similar equipment needed some other means of rectification.

SOLID-STATE DIODES

While the vacuum rectifier was having its day, the crystal diode held a relatively insignificant position in the electronic world. After World War II, a solid-state diode using selenium as its critical element was widely used in battery chargers and other low-voltage equipment. The selenium rectifier worked, but was very temperature sensitive. It had large cooling fins, but even so, burnouts were common.

It wasn't until about 20 years ago that the development of semiconductor electronics introduced the silicon diode, which presently holds the limelight as a rectifier. It is compact in size, operates at a low voltage, and can carry large amounts of current.

There are a number of ways to connect rectifier diodes, and the choice is largely up to you, although you should consider the particular devices you want to use, and what you want to do with the dc once you have it.

CIRCUITS AND SYMBOLS

Before we get into circuits for diodes, however, let's review some of the basic electrical and electronic symbols used in diagrams, as shown in Fig. 8-7.

When any two devices are connected together, the symbols are joined with a line. The line may represent a wire, or it may simply indicate a connection. In showing an electrical or electronic circuit, the symbols are laid out on the paper in a position that is both graphically pleasing and convenient to show the circuit accurately, and lines are drawn to show the connections.

THE HALF-WAVE RECTIFIER CIRCUIT

The simplest of all rectifying devices consists of a diode in series with the load, shown in Fig. 8-8A.

The half-wave rectifier allows current to flow in one direction, but not in the other. This is accomplished simply by blocking one half of the ac cycle. It works, but there is considerable space between the dc pulses, as shown earlier in Fig. 8-6. Half of the available power is lost. Consequently, this circuit is only used in devices where efficiency is not important.

When the power source is a transformer secondary that hap-

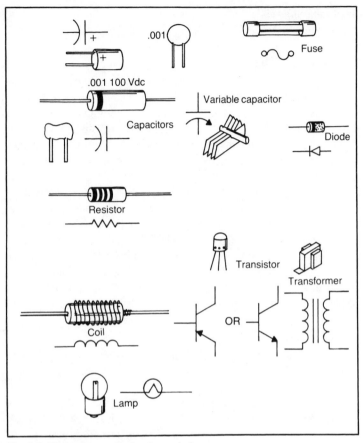

Fig. 8-7. The common electronic symbols.

pens to have a center tap, you can make that center tap the common connection and use each half as a separate ac source. Then, by connecting the two halves in parallel through individual half-wave rectifier diodes, you get what is commonly called the full-wave center-tap rectifier circuit.

This circuit uses each half of the transformer secondary alternately, thereby using both halves of the ac cycle. Since the two halves of the secondary reference to the center tap rather than to each other, the rectified voltage available is only that from the center to either end of the secondary, that is, one half the total secondary voltage. Since the complete ac cycle is used, however, the full power is available at the output. See Fig. 8-9. It may be only half the voltage, but now twice the current can be delivered.

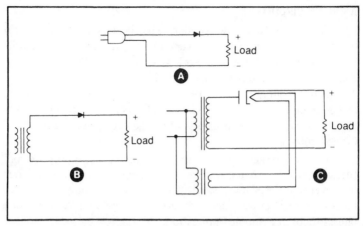

Fig. 8-8. Here are three half-wave rectifier circuits. The first two use a solid-state diode (A and B); the one at C uses a vacuum tube. Note that an extra transformer is needed to light the filament. An older tube that had no cathode would have the output taken from one of the filament wires.

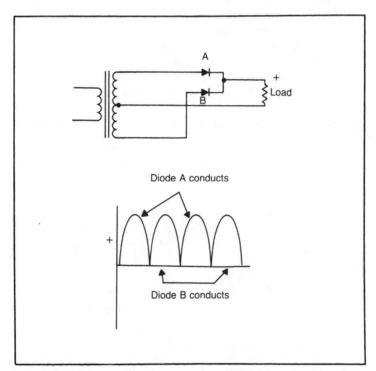

Fig. 8-9. A full-wave rectifier, and its output. The output voltage was half the transformer voltage. The name for this kind of circuit is full-wave center-tapped.

A BRIDGE CIRCUIT

The reader may now wonder, what do you do if there is no center tap in the transformer secondary, but you still want full-wave rectification? You can work around that problem simply by connecting two diodes to each end of the transformer secondary. The diodes from either end are connected opposite to one another and go to either end of the dc circuit. This is called a *bridge* circuit, Fig. 8-10, and uses both halves of the ac cycle, at the full transformer voltage.

The bridge rectifier is the most common general-purpose rectifier circuit in use today. Rectifier modules consisting of four diodes encapsulated into a single package are commonly available at most electronic stores, and they are quite inexpensive. Let's use one of them in a simple application.

Battery Trickle Charger

Materials you need are:

18-volt bell transformer
4 rectifier diodes, 15 PIV, 2 amp, or
1 rectifier bridge, same rating
Type 93 "hi-intensity" bulb
12 feet No. 10 insulated, stranded wire
2 battery clips
Metal or plastic box about the size of a lunch box (which will do nicely)!

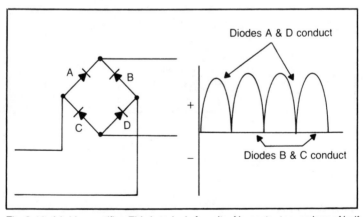

Fig. 8-10. A bridge rectifier. This is today's favorite. No center tap, no loss of half the power, and it uses the complete ac cycle.

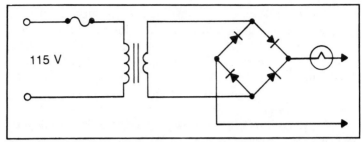

Fig. 8-11. A simple battery charger. The lamp limits current in the event of an accidental short circuit. Can you build it all the way using only the schematic diagram?

Power cord with plug
½ amp "slo-blow" electronic equipment fuse
Fuse holder (Get it at the same place as the fuse)
⅜-inch rubber grommet

Mount the transformer securely into the box, with the switch and fuse holder nearby. Drill a hole to receive the grommet, and pass the power cord through. Knot the power cord to provide strain relief.

The rectifier bridge or the diodes (whichever you use) can be mounted in any of several ways. Electronic hobbyists often use a material called "Perfboard" which is a phenolic (plastic) material

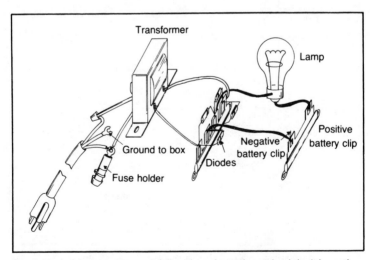

Fig. 8-12. Just in case you can't follow the schematic yet (and don't be embarrased if you can't), here's a pictorial of the battery-charger hookup.

with a grid of small holes drilled through it. Parts can be arranged on this material, and the board mounted in the box with stand-off bushings.

An alternative method is to use a multi-contact, phenolic terminal strip. This is best applied if you are using separate diodes rather than a ready-made bridge-rectifier module. A wiring diagram and parts placement drawing are shown in Figs. 8-11 and 8-12.

This charger should be able to put 1 to 2 amps into the battery, depending on the transformer used. When you use the charger, be sure the positive clip goes to the positive post on the battery.

You can produce a more powerful charger simply by using diodes rated at a heavier current, and a larger transformer. The transformer voltage should be no more than 18 to 20 volts, and no more than 5 amps or so.

FILTERING THE DC

As we have mentioned, the voltage output from this charger consists of 120 pulses per second. The ripple is intolerable for most electronic devices, although it is of no great consequence in a battery charger. To power sensitive equipment, some method of smoothing out the ripple is needed.

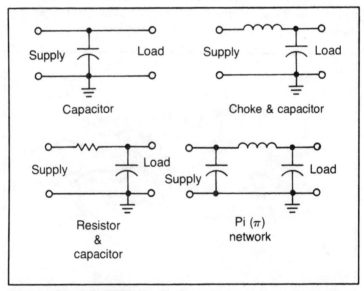

Fig. 8-13. Here are several circuits to filter the ripple out of the rectified dc. You may find that the circuits cause the output voltage to go up. This is because filtering enables you to use more of the energy than you could before.

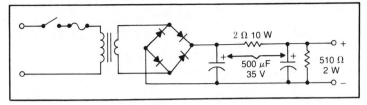

Fig. 8-14. Filtering your battery charger.

We said in an earlier chapter that alternating current flows easily through a capacitor, and is impeded by an inductance. The ripple output of a rectifier is an ac component, and if this can be prevented from reaching the equipment, nothing but smooth dc will be left.

Power supplies in the early days of electronics used an inductance in series with the load. The inductance let the direct current flow through quite easily, but the alternating-current component was greatly reduced.

The next step was to connect a capacitor right across the rectifier output. There are two theories as to how this worked. One is that the capacitor represents a short circuit to the alternating-current component. Personally, I found the other theory easier to relate to: The capacitor charges during the voltage pulses, and discharges between them, thus filling the gap. Regardless of which explanation you find easier to understand, the point is that it works. The present state of the art is such that many power supplies either use a resistor in place of the inductor, or omit it entirely. This is possible only in the lower-voltage supplies where large values of capacitance are possible in relatively small packages. Figure 8-13 shows some filter circuits.

The battery charger you just built can be converted easily into a low-voltage power supply simply by replacing the output leads with terminals, and connecting a large capacitor across the rectifier output, Fig. 8-14. You will note, however, that the filtered output voltage is considerably more than expected. When a capacitor is connected at the output of the rectifier, it charges to the peak voltage of the ac input voltage. The 18 volts at which the transformer is rated is the *rms* voltage. The *peak* voltage is 1.4 times the rms. Consequently, the dc output voltage (with no load) will be about 25 volts.

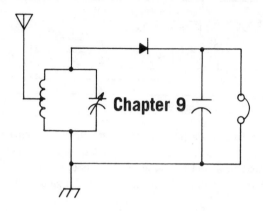

Vacuum Tubes and Transistors

The chapters that follow can only lightly gloss over their particular subjects. This is because the state of the electrical arts is changing so rapidly, significant changes can occur faster than books can be written and printed. We will confine our efforts, therefore, to explaining the basic, developmental principles involved in each field of coverage, and recommend the reader to other excellent TAB books on the subject.

THE VACUUM TUBE

In the very early years of this century, a man named Lee DeForest undertook the task of developing a new form of radio detector. His theory was that radio signals could be made to ionize gasses within an incandescent lamp, thereby changing the resistance between the filament and an adjacent electrode.

While his original theory was not quite like the actual operation of the final product, his work resulted in the birth of one of the world's largest and most beneficial industries. All industries are, in one way or another, touched by the results of DeForest's work, even though the invention itself is now becoming obsolete.

There is little point here in describing the development of the vacuum tube step by step, other books cover that part of history. Let's confine our discussion to the final product.

DeForest knew he was on to something, so he performed his work at night, in secret. The device itself was housed in a small, wooden box with the wires leading into it, and batteries connected to it. When DeForest finally got it working, he handed the earphone to his assistant. "My God, hear those signals!" the man exclaimed, "What have you got in that box, Doc?" His astonishment was well justified. The signals were five to ten times louder than what might normally be expected from any other known radio detector. De-Forest had developed a device that could amplify electric signals.

It was very similar to the vacuum rectifier described in the last chapter, except that a third electrode was added. This electrode consisted of wire grid surrounding the filament, see Fig. 9-1.

Here's how it works. When the filament (nowadays the cathode) is heated, it throws off electrons which, in turn, are attracted to the positively charged plate. If a slight negative charge is applied to the grid, it repels the electrons, disturbing their flow to the plate. Any small variations in the voltage applied to the grid result in large variations of the electron flow. This, in turn, produces large variations in the plate voltage and current that follow exactly the variations in the grid voltage, see Fig. 9-2.

The net result is that the tube adds a considerable amount of power to the signal. Maybe the device is not so amazing today, but it

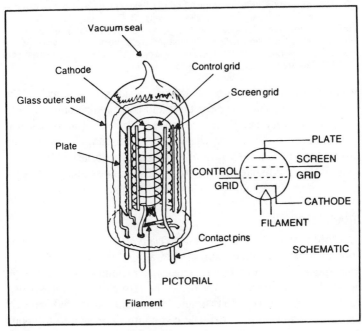

Fig. 9-1. Inside view of a typical vacuum tube.

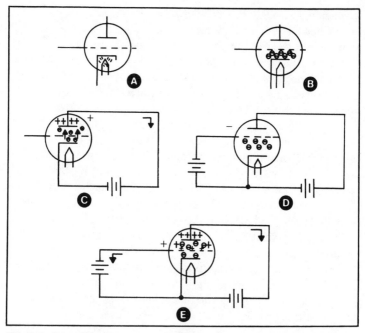

Fig. 9-2. Operation of a vacuum tube. (A) Filament heats up the cathode (Cathodes were not used in the earliest tubes). (B) Electrons are driven off the cathode by the heat. (C) A positive charge on the anode (plate) will attract the electrons allowing current to flow. (D) A negative charge on the grid will repel the electrons, preventing them from reaching the plate. (E) A positive charge on the grid accelerates the electrons on their way to the plate.

was revolutionary for its time. It was a major step forward for the telephone companies, and for the radio communications industry. The vacuum tube is still around today, in spite of claims that it is becoming obsolete. It has changed its form and additional electrodes have been added. Literally thousands of variations are available. While solid-state devices are rapidly gaining ground, tubes are still used in some radio communications equipment.

TRANSISTORS

About the time that television receivers were becoming common in homes, the electronic world was set humming by tales of a tiny speck of metal that could make the vacuum tube obsolete. It was called the transistor, a word coined from "transfer resistor," and while they were at first fragile (electrically speaking) and very limited in application, they have come into their own to the point that a computer which once occupied a large room and contained upwards of 18,000 tubes can now be replaced by a better one that sits on a desk top.

Transistors do not need to be enclosed in a glass bulb, they do not require filament power just to make them operational, and since the individual parts are in one solid mass instead of being mounted inside the bulb, they are mechanically more durable. Electrically speaking, tubes still hold the edge in tolerance of operator error. However, while transistors are less forgiving than tubes when it comes to electrical abuse, their advantages far outweigh that one disadvantage.

All the years of development of electronic circuits using tubes were by no means wasted. Many of the circuits can be directly applied to transistors with minor variations.

TRANSISTOR TYPES

There are several kinds of transistors in use. The most common are the junction and the field-effect (FET) transistors. Before either can be understood, it is necessary to explain the kinds of material they are made of.

There are certain metals which, under controlled conditions, can be either conductors or insulators. Such metals are called semiconductors. The two most widely used semiconductors are Germanium and Silicon. In order to make these suitable for transistor use, they must be purified to an almost unbelievable degree. One party described the purity as being equivalent to one grain of salt in ten carloads of sugar.

Once the semiconductor metal is pure enough, controlled amounts of specific impurities are deliberately added to determine the electrical characteristics of the metal. Two types of impurities are added: one that puts a surplus of *electrons* into the molecular structure, and the other which causes a deficiency of electrons (or a surplus of *holes*). Material with a surplus of electrons is called *P-type;* material with a surplus of holes is called *N-type*. See Fig. 9-3.

The Junction Transistor

The *junction* transistor, which has long been the workhorse in the solid-state field, can best be described as a miniature sandwich. There are two kinds, PNP and NPN. Figure 9-4 describes their construction.

Their theory of operation is far too complicated to express in simple terms. The main difference between the two is that the

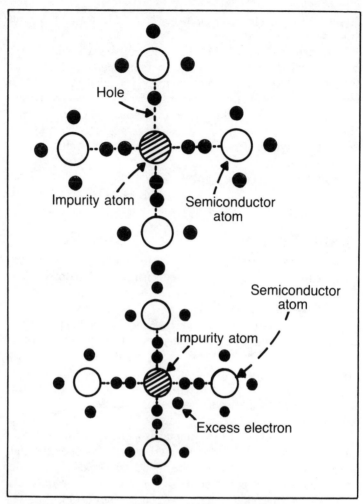

Fig. 9-3. Atomic structure of P-type and N-type germanium.

power connections for one is opposite those for the other. The three terminals are called *base*, *collector*, and *emitter*. Their connection into a circuit will be described shortly.

The FET

A *field-effect transistor* (FET) has terminals similar to those of a junction transistor, but the operation and construction are very different. The terminal similar to the base of a junction transistor is called the *gate*. The other two are, respectively, the *source* and the

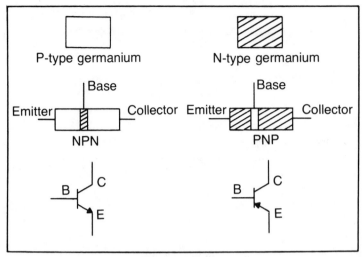

Fig. 9-4. Construction of junction transistors.

drain. See Fig. 9-5. Electron flow between the source and the drain is affected by voltage applied to the gate, and that effect results in amplification.

Just as junction transistors are expressed as either PNP or NPN field-effect transistors can be either N-channel or P-channel.

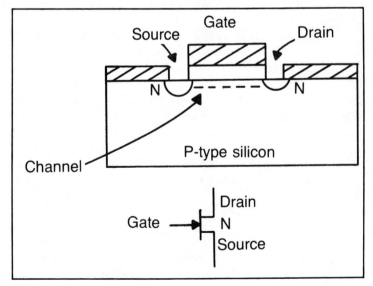

Fig. 9-5. Field-effect transistor.

We will describe the construction of an N-channel unit, and you need only remember that, in a P-channel, all the elements are of the opposite kind.

An N-channel FET has a basic material of P-type silicon, into which is diffused (melted) two N-type semiconductor areas. These form the source and drain elements of the transistor.

On top of this a very thin silicon-oxide (glass) insulating layer is diffused. Over that is diffused a third element called the gate. The area beneath the gate is called the *channel*. Electrons flow between the source and drain if the drain is made positive with respect to the source. The level and polarity of the voltage applied to the gate determines the conductivity of the channel. Consequently, an alternating-current signal on the gate will cause the channel current to vary, and the variations will be an amplified version of the signal on the gate.

Operationally speaking, a transistor amplifies current while a tube amplifies voltage. Except for that, the electrical principles for connecting either one are similar.

AMPLIFIER TECHNIQUES

There are a number of ways to connect a signal into a tube or transistor to amplify it. The most common is the one in which the signal inputs on the grid of the tube, or on the base or gate of the transistor. The amplified signal appears at the plate of the tube, or at the collector or drain of the transistor, see Fig. 9-6.

In order to operate, a vacuum tube must first have the filament lighted (cathode hot). Second, there must be a fairly high dc voltage applied to the plate with respect to the cathode. (In some tubes, the filament is the cathode; in others, the filament heats a chemically-treated sleeve, which is the cathode.) Finally, a slight negative voltage, called the bias, must be applied to the grid. The bias voltage ensures that the plate current variations follow exactly the variations in grid-signal voltage, thereby preventing distortion. If the tube has more than one grid, the second grid usually has a positive dc voltage applied, while the third grid is usually connected to the cathode. See Fig. 9-7.

On the other hand, a transistor is biased by current rather than voltage. Furthermore, the polarity of that voltage depends on whether the transistor is PNP or NPN. In either case, the filament-voltage supply is unnecessary, since there is no filament.

The operating supply is connected between the collector and the emitter. In order to turn the transistor on, and to ensure the

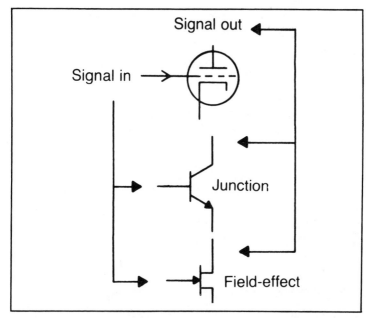

Fig. 9-6. Analogy between a vacuum tube and two types of transistors.

proper operating characteristics, a slight current is caused to flow between the base and the emitter.

With PNP transistor, the operating supply must put the negative voltage on the collector with respect to the emitter. Then a slight negative voltage is applied to the base to cause base current flow. This voltage is easily supplied by means of a resistance voltage divider across the collector supply. Connections for an NPN transistor are identical, except that the polarity is reversed. Connections for both types are shown in Fig. 9-8.

Once the proper power connections are made to the amplifying device, it is simply a matter of feeding the appropriate signals in and out. They are usually coupled in through a capacitor or transformer (which isolates the signal-handling circuits from the direct-current power voltages.)

Circuits

Let's take a closer look at these. Even though vacuum tubes are obsolete, it is best that we begin with a simple vacuum-tube circuit. This will help to understand solid-state circuits.

The signal is fed into the grid of the tube through a capacitor which isolates the grid from the signal source. A high resistance

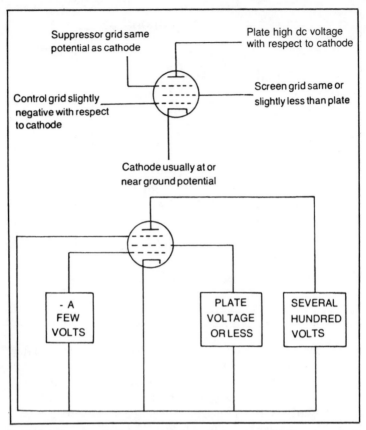

Fig. 9-7. Power supply requirements for a vacuum tube amplifier.

from grid to ground sets the input impedance and bleeds off charges that would otherwise accumulate due to the internal operating conditions of the tube.

The grid must necessarily be held to a specific voltage, slightly more negative than the cathode. In older circuits, this used to be done by a separate battery. There is an easier way: simply insert a small resistance in series with the cathode. Since current is flowing through the cathode, a voltage drop results, making the cathode slightly more positive than ground. The grid is essentially at ground level, therefore negative with respect to the cathode.

The amplified signal is recovered from the plate of the tube through a capacitor which isolates the subsequent circuits from the relatively high plate voltage. Plate power is supplied through a series resistor that limits the current and sets the output charac-

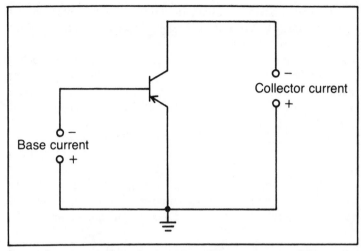

Fig. 9-8. Power requirements for a transistor. The one shown is a PNP type. For NPN type, use opposite polarity of all current supplies.

teristics of the tube. This resistor is usually bypassed for ac through a capacitor at the power-supply end, see Fig. 9-9.

Now let's look at the same device using a junction transistor instead of a vacuum tube. As before, the signal is fed in through a capacitor. The capacitor has a much higher value, because transistors are generally low-impedance devices.

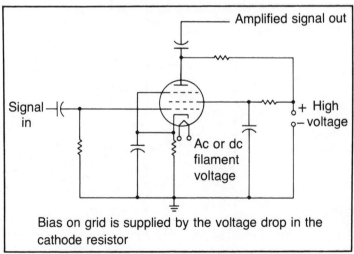

Fig. 9-9. A vacuum-tube amplifier circuit. This tube has three grids, and is called a pentode. Note how the voltage is supplied to each.

Instead of having a voltage on the grid, we now must force a small amount of electron flow through the base from the emitter. The amount of current is set by a resistor in series with the emitter. In a junction transistor, the base voltage sets itself to approximately 0.7 volt higher than the emitter voltage. The value of the emitter resistance is calculated with these considerations together with the rated base current of the transistor. It's a simple application of Ohm's-law equations.

The collector voltage is set to a point about midway between the supply voltage and the base voltage, by means of a series resistor.

One of the most important audio-amplifier circuits is the one shown in Fig. 9-10, called the emitter follower. The signal is applied to the base, and recovered in series with the emitter. The

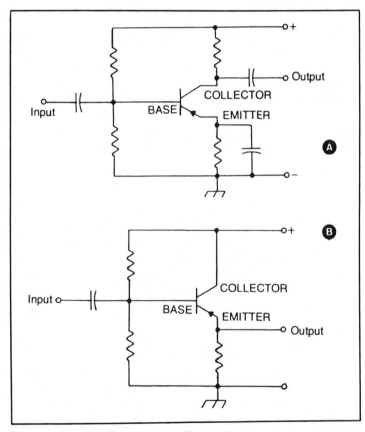

Fig. 9-10. Two kinds of transistor amplifier circuits.

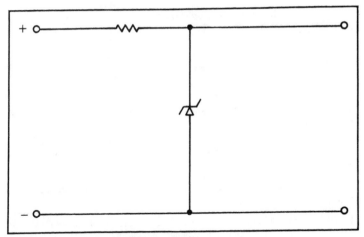

Fig. 9-11. A Zener-diode regulator circuit.

advantage of this circuit is that it has a high input impedance and a low output impedance. This means that it can couple directly to a low-resistance device, such as a loudspeaker.

The emitter follower circuit is also used to regulate the dc output of a power supply. It is easy to understand the operation of this circuit. Remember that the emitter voltage of a silicon junction transistor follows the base voltage, but about 0.7 volt lower. The power supply output is connected to the collector of the transistor, and the load to the emitter. The base voltage is then set by means of a Zener diode.

A Zener diode is a special device intentionally designed to break down (conduct) at a selected reverse voltage. It is normally connected into the circuit in reverse, and if the voltage across which it is connected rises high enough, the diode will conduct. When the voltage drops below the conduction point, the diode then turns off. A resistor in series with the diode limits its current flow. When conducting, the diode will draw only that current required for the resistor to drop the voltage back to the breakdown point. See Fig. 9-11.

The result of this is that the voltage across the Zener diode holds relatively constant under varying load conditions.

If the Zener diode is connected to the base of the emitter-follower circuit so that it sets the base voltage to a constant value, Fig. 9-12, the emitter voltage will follow, remaining constant under a much wider range of load conditions than would be possible without the transistor.

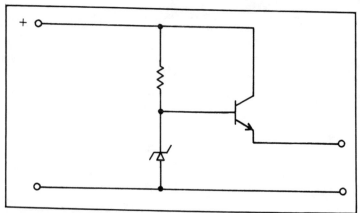

Fig. 9-12. A Zener diode regulator reinforced by a transistor. This permits the use of a much lower power diode, and smaller capacitors.

The circuits just described are only a few of the applications of transistors in electrical and audio work. To describe them all would take a much bigger book than this. There are, however, a great many TAB books on the subject, and we recommend them for further reading.

We are now about to try a few experiments, but first, in order to do those experiments, you may want to make an experimenter's *breadboard*.

The term "breadboard" comes to us from some early pioneers in electronics who built their experimental circuits on a base made from a cutting board borrowed from the kitchen. This offered a method of mounting that left all components accessible for testing and experimenting.

In the research labs and factories of today, a "breadboard" layout of a new circuit is built on a piece of perforated board such as was mentioned in the last chapter for assembling the power supply. Small clips (called flea clips) are inserted into the holes and the components soldered to them. When you use components that are relatively heavy, such as a vacuum tube or transformer, you may prefer to use a wood base.

VACUUM-TUBE AMPLIFIER EXPERIMENT

You'll need some equipment and materials for these experiments, such as:

12AU7 tube
12-pin tube socket

12-volt filament transformer
2 resistors, 47,000 ohms, ½ watt
1 resistor, 1,000 ohms, 1 watt
1 capacitor, 5 μF, 25 volts
2 capacitors, 0.05 μF, 200 volts
Power supply, approximately 100 volts dc
Phono pickup, microphone, or other signal source
Experimenter breadboard
Headphone

CAUTION: Dangerous voltages are used in this experiment. Adult supervision is recommended.

The small numbers on the schematic diagram (Fig. 9-13) indicate the particular pin numbers to which connections must be made in order to reach the elements shown in the circuit. Sometimes the corresponding numbers may be indicated on the tube socket. If not, here is how to find them: Hold the socket upside down. Note where the space is in the circle of pins. Position that space toward you. Pin number 1 is the left-hand one to the space. Then you simply count in a clockwise direction. See Fig. 9-14.

Connect the parts into the circuit as shown, and apply heater voltage to the tube using the filament transformer. Allow about a minute for the tube to warm up, then apply about 100 volts high voltage.

Feed a signal into the grid using either the microphone or some other signal source. Listen to the input signal through the earphone.

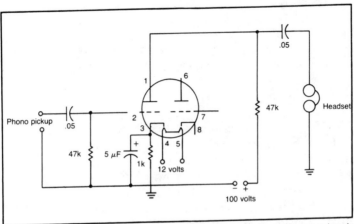

Fig. 9-13. A simple vacuum-tube amplifier. The tube used is really two tubes in one bottle. Only one side is used.

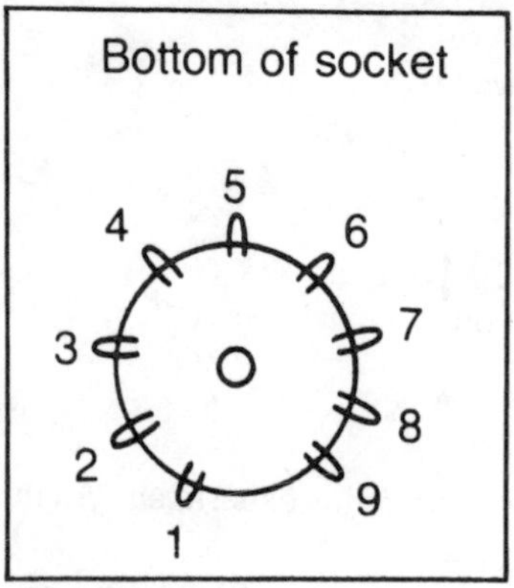

Fig. 9-14. Pin numbering on a vacuum-tube socket. This is as the socket is viewed from the bottom.

Now move the earphone to the output coupling capacitor. Is the signal louder?

If you want, you can add a second stage equal to the first, Fig. 9-15. See how much amplification you then have.

TRANSISTOR AUDIO AMPLIFIER

Materials and equipment needed to make a transistor audio amplifier include:

PNP audio junction transistor
NPN audio junction transistor
3 capacitors, 0.1 μF, 15 volts or more
Resistor, 120 ohms, ¼ watt
Resistor, 300 ohms, ¼ watt
Resistor, 680 ohms, ¼ watt
Resistor, 3300 ohms, ¼ watt
9-volt transistor battery and battery connector
Signal source, same as for vacuum-tube amplifier
Earphone
Experimenter breadboard

The schematic symbol and battery connections shown in Fig. 9-16 are those for an amplifier using a NPN transistor. If a PNP transistor is used, the circuit remains the same, but the battery connections are opposite those shown. Also, if the capacitors used have polarity markings, turn them around.

Duplicate the tests you made with the vacuum-tube amplifier. First listen through the earphone to the loudness of the signal directly from the source. Then apply the signal to the amplifier, and

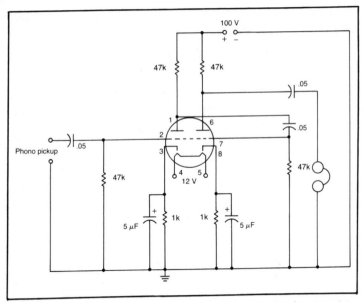

Fig. 9-15. If you want to make full use of the tube, use both halves for a two-stage amplifier like this.

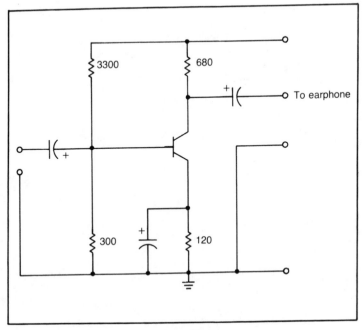

Fig. 9-16. A simple transistor audio amplifier.

153

connect the battery. For such a tiny thing as this, the transistor has a surprising amount of gain! You will find that the opposite types (PNP or NPN) work equally well.

It would be foolish to go much deeper into transistor circuitry here. There are a great many TAB books, written at all levels, on that subject. Moreover, the Radio Shack and numerous other manufacturers offer excellent beginner experimenter kits to get one started into virtually any phase of electronic experimentation.

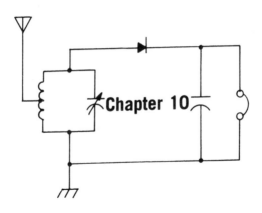

Radio Communication

Of all the beneficial uses of electric phenomena, few have had a greater impact on our society in general than radio communication. By it, we learn of important events, no matter where in the world they may occur, in a matter of minutes; we discover the ideas and customs of our fellow man in other cultures ; and we can, in times of great need, summon the assistance of our near neighbors.

In civilized countries, radio communication affects the lives of practically every household. In the lesser developed nations, it still reaches far into the wilderness to encourage learning and evaluate the general level of all peoples.

Nobody knows for sure exactly what radio waves are, although there are some very sound theories. What is known is that, when an alternating voltage is imposed at a certain frequency onto an antenna, radio waves are produced at that frequency, and travel outward at the speed of light. (In fact, light waves and radio waves are of the same nature, but of different frequency.)

Radio waves are produced by alternating currents above about ten to fifteen thousand cycles per second. When a radio wave strikes any electric conductor, it induces an alternating current in that conductor at the same frequency as the wave.

Long before any kind of radio waves were produced, they were known to exist. In fact, as early as 1865, James Clark Maxwell mathematically proved their existence. After Maxwell's startling publication, however, over 22 years were to pass before physical proof was accomplished.

In 1887, Heinrich Hertz performed a carefully calculated experiment to prove Maxwell's discoveries. He discharged a Leyden jar through a loop of wire. Nearby, a similarly-sized loop with a gap in it produced a spark. See Fig. 10-1. In later experiments, Hertz was to prove that these mysterious waves were similar in nature to light waves—they could be focused, reflected, and refracted. He was, however, far ahead of his time. Some of his experiments used frequencies in the microwave region, and it was to be many years before those frequencies could be put to practical use.

EARLY SYSTEMS

Within a decade, sensitive detectors were being produced, and Sir Oliver Lodge outlined a complete wireless system. In Russia, Popoff put a telegraph system into actual operation. The authorities felt that such communication was "subversive" and banned it. It wasn't until Marconi proved the commercial feasibility of radio telegraph that the system became popular.

One of the more important of Hertz's discoveries was that these mysterious waves could be tuned. At first, he didn't quite understand what was happening. What he did know was that the loops on the transmitter and on the receiver had to be about the same size, a condition that he called "syntony." In 1898, Lodge

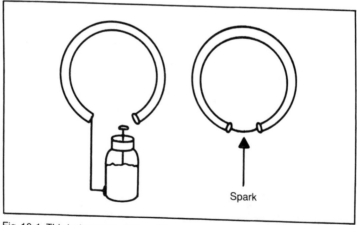

Spark

Fig. 10-1. This is the general idea of the apparatus used by Hertz to demonstrate the existence of radio waves. The Leyden jar was first charged. Then it was discharged through the loop. On the other side of the room, another loop of similar size would produce a spark at the instant the Leyden jar was discharged.

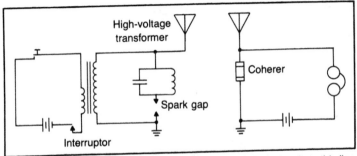

Fig. 10-2. Once radiotelegraph was established, it worked more along this line. The spark gap would set up radio frequency oscillations in the transmitting antenna, which the receiver's coherer would pick up. Not shown is a vibrator which would decohere the detector and get it ready for the next signal.

perfected the coil-and-capacitor tuning system, and three years later Marconi made his first transatlantic radio contact.

Marconi used a device similar to an automobile ignition coil to produce radio signals, and for a detector, he used a small tube filled with metal filings, called a *coherer*, Fig. 10-2. The principle was that, when radio signals passed through the filings, they would cohere, or stick together, making the internal resistance of the device much less. This would allow a local battery to close a relay, ringing a bell. At the same time as the bell rang, a vibrator would loosen the filings in the coherer, preparing it for the next radio impulse. It was, at best, a very inefficient device by today's standards.

Then came the crystal diode, discussed in a previous chapter, and DeForest's vacuum tube. Much of the early development and experimentation was done by amateurs. On the professional side, Fessenden produced the first voice-modulated broadcasts in 1900 and 1906. Radio operators all over the Eastern seaboard of the United States were astounded. One ship's radio operator entered into the log that he heard "angels singing."

Radio really came into its own, however, after World War I. In 1920, five American stations began commercial broadcasting, and within two years, there were over 500 on the air.

Since the invention of the vacuum tube and its related circuits, the means of producing radio waves has changed only slightly, except for the transition from vacuum tubes to transistors. Methods of modulation and very-high-frequency techniques have changed radically, while still using many of the earlier ideas.

The thing that produces the high-frequency alternating current which the antenna converts into radio waves is called an oscillator. In past years, an oscillator made from a high-power tube made a simple, one-tube transmitter. These were very popular among beginners on the Amateur bands for code transmission. There were even some simple, one-tube voice transmitters, Fig. 10-3, but they simply don't make the grade for today's operating techniques.

VOICES

The earliest, and for many years the main method of transmitting human voice involved the use of an audio amplifier equal in power to the output amplifier of the transmitter, which imposed the audio signal directly onto the radio-frequency output of the transmitter. This is still used in some CB transceivers. It results in what is called a double-sideband, AM signal.

Sidebands result from the natural mixing of the audio and radio frequencies. They are caused by a law of nature that says that, when two signals of different frequencies are mixed together, they produce signals at the sums and the difference of the two. Applying that rule, if an audio frequency of 1000 Hz was mixed with a million hertz (1 MHz) of radio frequency (a hertz is one cycle per second), *sidebands* would be produced at 1,000 hertz above and below the one-million-hertz signal. That is, the sidebands would occur at 999,000 and 1,001,000 hertz. See Fig. 10-4.

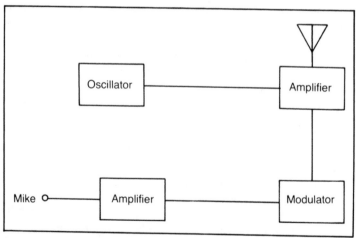

Fig. 10-3. Today's radio transmitter (at least the AM transmitter of the last decade) consists of an oscillator which generates the radio signal, and an amplifier to give it the power. The voice was first amplified, then fed to a modulator which put it onto the radio signal.

158

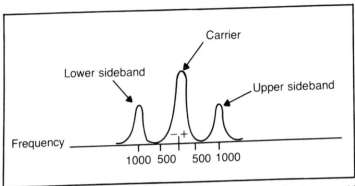

Fig. 10-4. If you made a graph of the frequencies being transmitted, you would find that a carrier wave and two sidebands were present.

You will notice from the illustration that an AM double-sideband signal consists of a carrier wave and the sidebands. If the signal is being modulated with a human voice, the sidebands will fluctuate in amplitude and frequency within the limits of the carrier plus and minus the voice spectrum. In a fully-modulated signal, the sidebands contain half the transmitter power and the carrier the other half. This is a tremendous waste. The only function of the carrier is to mix with the sidebands in the receiver to recover the voice.

Shortly before World War II, a means was developed to eliminate the carrier and one of the sidebands, putting the entire transmitter power into the remaining sideband. With this technique, all the available transmitter power is used to transmit the information. A carrier is inserted in the receiver at a much lower power, and the voice can then be recovered. This technique is known as suppressed-carrier, single-sideband transmission, Fig. 10-5. Besides using only half as much of the radio spectrum, it has the effect of a two- or three-to-one increase in power over an AM signal of the same power.

We have so far confined our discussion to transmitters, as they are generally the simpler devices in radio communication. At least, they were until single sideband communication came into its own. It is relatively easy to put a signal onto the air. Recovering that signal and sorting it out from among all the others is another matter entirely.

When a radio wave strikes a conductor, it induces minute alternating currents of the same frequency as the wave in the conductor. Now, there are thousands of different waves striking

159

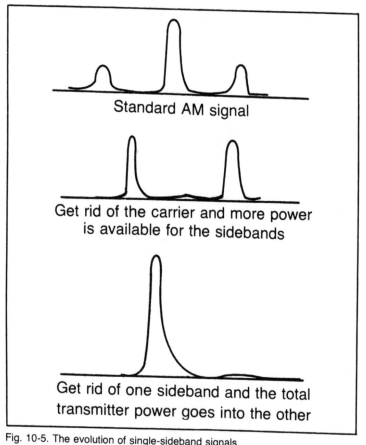

Standard AM signal

Get rid of the carrier and more power
is available for the sidebands

Get rid of one sideband and the total
transmitter power goes into the other

Fig. 10-5. The evolution of single-sideband signals.

every antenna constantly. Before detecting them, we must first sort out the wanted one.

RESONANCE

In order to better understand the process of electrically sorting signals of different frequencies, we need first understand the mechanical parallel to the phenomenon. Every object that is made to vibrate has what is called a natural period. That is, a frequency at which it vibrates much more easily than any other. That frequency is referred to as the *resonant period* of the object, and is affected by a number of factors.

A good example is the string in a musical instrument. Its natural resonant period depends on such factors as length, thickness, elasticity, and tension. Vary any of these, and the frequency at which the string will vibrate changes.

Resonances were long a major bug-a-boo in the development of microphones, earphones, speakers, and all devices associated with the reproduction of sound. As the art progressed, scientists learned how to avoid those resonances, and when unavoidable, to use them to an advantage.

All this has been paralleled in electrical considerations. Just as mechanical devices have mechanical resonances, electrical devices have natural resonances at which frequencies alternating current functions with the maximum efficiency. It is by the careful use of such resonances that we are able to separate signals in receivers.

The two factors having the greatest influence on electrical resonance are inductance and capacitance. Except at the very high frequencies, these two factors are relatively easy to control, not only through good layout but through use of the correct components as well.

The handiest thing about inductors and capacitors is that they are electrically opposite to one another. Certain factors in their opposition to the flow of alternating current (a factor dependent on frequency) are positive in an inductor and negative (relatively speaking) in a capacitor. Thus one can effectively cancel out the other.

The inductive and capacitance factor that opposes the flow of alternating current (to an extent partly determined by frequency) is called reactance. Inductive reactance varies directly with frequency; capacitive reactance varies inversely with frequency. That is, as frequency increases, inductive reactance becomes greater and capacitive reactance becomes less. See Fig. 10-6.

It stands to reason, then, that in a circuit having both inductive and capacitive components, there must be a frequency at which inductive and capacitive reactances are equal. At that frequency, and that frequency alone, the two cancel one another out entirely and the only opposition to the signal is the pure resistance. Therefore, the circuit will show a far greater efficiency at that frequency than at any other. It is said to be *resonant* to that frequency.

Let's look at that a little more closely. If a coil and a capacitor were connected in series, Fig. 10-7, the coil would offer greater opposition to high frequencies than at low frequencies. The exact amount of opposition would depend on its inductance as well as on the frequency. The capacitor would offer greater opposition to

161

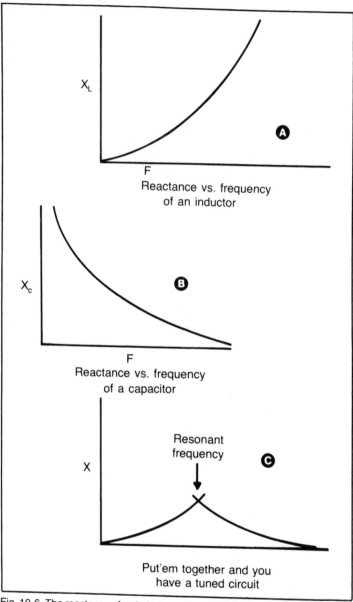

Reactance vs. frequency
of an inductor

Reactance vs. frequency
of a capacitor

Resonant
frequency

Put'em together and you
have a tuned circuit

Fig. 10-6. The reactance of an inductor varies with frequency (A). The higher the frequency, the higher the reactance. With a capacitor, it's the other way around (B). The higher the frequency, the lower the reactance. This means that there is one particular frequency that can flow best through an inductance/capacitance combination. That frequency is called the *resonant*. At resonance, the two reactances cancel one another (C), and only resistance is left to be dealt with.

lower frequencies. The exact amount of opposition would depend on the capacitance as well as on the frequency.

If the values of the two components are fixed at specific amounts of inductance and capacitance and remain unchanged as the frequency is varied, there would be a point where one cancelled out the other and only the pure resistance of the wire would be left. Higher frequencies would be opposed by the inductor; lower frequencies would be opposed by the capacitor. This point is known as the natural resonant frequency of the circuit.

Now picture the two components in parallel with one another, connected across the signal source, Fig. 10-8. Again let us assume that the values are so selected as to be resonant at the signal frequency. The inductor tends to short out the lower frequencies; the capacitor tends to short out the higher frequencies. This is really oversimplifying what actually happens, but will do for what we have to say here.

At the resonant frequency, a parallel tuned circuit offers maximum opposition. There are other factors, much too advanced to go into here, but let us simply say that a parallel tuned circuit also emphasizes its resonant frequency, much like an organ pipe does with sound. A parallel tuned circuit, connected across a signal

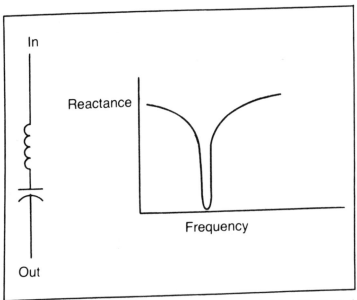

Fig. 10-7. The reactances of a series tuned circuit roughly follow this curve. A series tuned circuit lets the resonant frequency pass through easily.

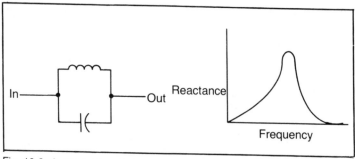

Fig. 10-8. A parallel tuned circuit behaves oppositely to a series circuit. The resonant frequency is blocked.

source, will actually show an apparent increase in the signal voltage. It is, therefore, the type most often used in radio detection. In most AM radios, the capacitor is varied to select the resonant frequency.

From this lengthy discussion of resonance, we have shown the way in which incoming signals are separated from one another. Once this is done, the next thing to do is to recover the information from the selected signal. A simple tuned system is connected as shown in Fig. 10-9.

Once the desired signal is isolated by the tuned circuit, information can be recovered by rectifying it in the case of an AM signal, or by mixing it with a carrier in the case of a single-sideband or Morse-code signal. FM signals are in an entirely different ball game and the means of detection are a bit too complex for this book.

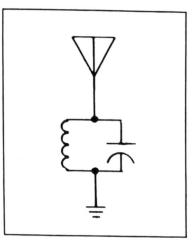

Fig. 10-9. The basic antenna/ ground circuit of a radio system uses a parallel tuned circuit. All unwanted frequencies are shorted to ground, and the desired frequency is left to be processed.

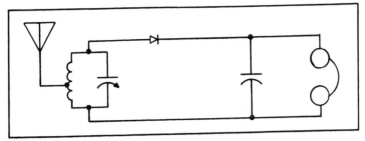

Fig. 10-10. The old crystal set of grandpa's age used this circuit. The old "catwhisker" detector was connected where we have the diode. You can build this set, and it will work. The variable capacitor has a value of 365 pF. The coil consists of No. 28 wire wound on a paper roller about 1¼-inch wide for a length of five inches. The fixed capacitor is about 0.001 μF.

RECEIVERS

The earliest receivers (once the coherer went out of style) consisted of a galena crystal with a thin wire called a catwhisker touching it. This arrangement acted like a diode. Substituting a diode for the crystal detector, we end up with the circuit shown in Fig. 10-10.

When the carrier and sidebands are selected by the tuned circuit, they then mix in the diode to form a radio-frequency signal with an amplitude that varies at an audio rate. This is then rectified by the diode to produce a dc voltage that fluctuates at an audio rate. This voltage is then fed into the earphone where it produces sound. The circuit is shown in Fig. 10-11.

The earliest transmissions were, of course, in code. However, they were still detectable with a diode detector because the trans-

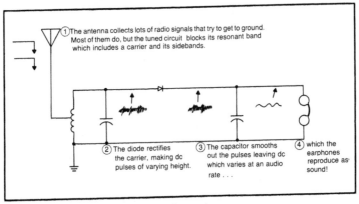

Fig. 10-11. Operation of a simple radio receiver.

mitter was a spark gap. The spark produced an interrupted signal which, once rectified, made a buzz in the earphone. A pure, steady-state signal would only have produced a tiny dc voltage, and nothing more than a click would have been heard in the earphone.

The earliest receivers were crystal detectors. Even the best of them had a tuned circuit that was primitive by today's standards. Even though there were relatively few stations on the air, interference was very bad. Perhaps that is why many of the early developments that improved the art of radio communication and helped prove the commercially developed circuits were accomplished by Amateur Radio experimenters. They had little choice but to get along with what they had, and to get the most out of it.

During the babyhood of radio, the typical Amateur transmitter consisted simply of a tuned spark gap. Simple as that may sound, it wasn't too far removed from the commercial gear. The receiver was simply a crystal detector. Figure 10-12 shows a typical one that was in the 1916 Boy Scout Handbook.

Please take our word for it that this thing worked, at least over relatively short distances. We do *not* advise any experimenter to build and operate it. While it did create a radio signal, that signal

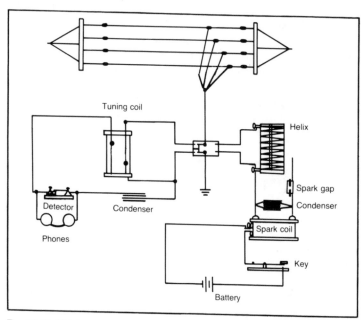

Fig. 10-12. This is a copy of the circuit of a two-way radio telegraph project in the 1916 Boy Scout Handbook. Do not try to put this onto the air!

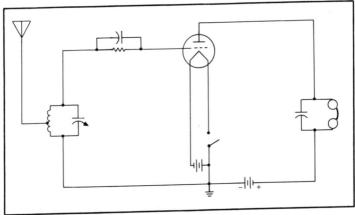

Fig. 10-13. The basic grid-leak detector. Tubes in those days had no separate cathode, and the circuit was wired as shown.

was filled with noise and harmonics. If you put a spark transmitter on the air today, you would have every Amateur, commercial, TV, and government radio station for miles around howling at your door—not to mention a neighborhood lynch mob and the FCC. You see, the signals it produces are not too clean, so there is some interference to everything from TV on down.

With the invention of the vacuum tube, the improvements in electronics grew tremendously. Hardly a year went by without a revolutionary new innovation. Vacuum-tube detectors applied the signal to the grid of the tube, and the grid was biased so as to rectify. That is, the tube only conducted during the positive half of the cycle. Consequently, the signal not only was detected, but amplified as well. Figure 10-13 shows the circuit of an early triode detector.

This detector was known as a grid-leak detector because of the resistance in series with the grid. The signal passed through the capacitor, but the dc charge that built up naturally on the grid was allowed to slowly leak off through the resistor. This kept the grid biased slightly negative so that signals could be rectified. If you wish to build this on your experimenter breadboard, the values given in the diagram will work. In the interest of space, we will give you the circuits without going into elaborate instructions, except as noted in the illustrations.

The setup with which DeForest first demonstrated the potential of the vacuum tube was probably a crystal detector with an audio amplifier coupled to it. A modern-day version of that setup is shown in Fig. 10-14.

167

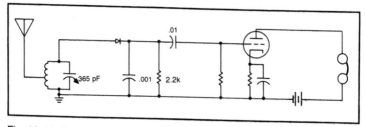

Fig. 10-14. Use the diode detector and the amplifier circuit from the previous chapter to make this version of the first amplified radio receiver.

The difference between this circuit and DeForest's is that the early versions used a transformer to couple the amplifier stage to the detector. If you add a couple of switch sections as shown in Fig. 10-15, you can then switch the amplifier in and out, demonstrating the amplifying ability of the tube.

Then somebody had the idea of adding an amplifier stage before the detector, Fig. 10-16. Since there were now two tuned circuits instead of one, the system had much sharper tuning, but there were some disadvantages.

A big breakthrough came when Major Edwin Armstrong discovered a method of feeding some of the amplified signal back to the input circuit. This brought about tremendous increases in gain, and made very sensitive, simple sets a reality. It was called the regenerative detector. See Fig. 10-17.

Up until World War II, the average Amateur receiver consisted of a regenerative detector with one or two stages of audio amplification. Then, shortly before the war, Major Armstrong devised another circuit called the superheterodyne. It operates on the principle that, when two signals of differing frequencies are mixed together, a new signal is produced.

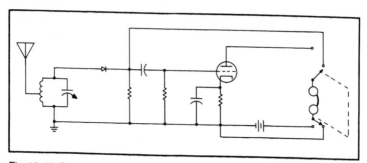

Fig. 10-15. Connect a double-pole, double-throw switch as shown. This will allow you to switch the amplifier in and out, demonstrating its effectiveness.

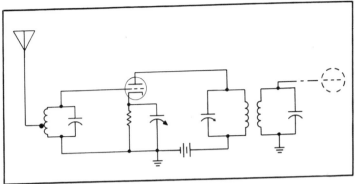

Fig. 10-16. A radio-frequency amplifier. This circuit is incomplete, and if you build it, it may oscillate.

Actually, there are two new signals. Their frequencies are the sum and the difference of the original two.

By using the new frequency that is lower than the original frequency, it is possible to use circuits that tune much sharper. Therefore, the whole radio had much sharper tuning, as well as being more sensitive than even the regenerative detector.

Let's look at the way a superheterodyne works. Instead of showing all the components, we'll represent each stage of the system as a box. Such a diagram is known as a "block diagram." See Fig. 10-18.

The signal may, depending on how expensive a receiver it is, pass through an amplifier (called an rf amplifier because it amplifies radio frequencies) into the signal mixer. There it combines with a signal from an oscillator to produce the intermediate frequency

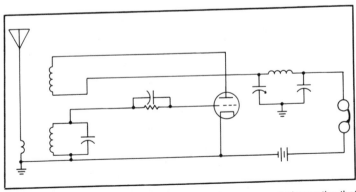

Fig. 10-17. Edwin Armstrong's regenerative receiver was a major innovation that became the favorite for Amateur receivers prior to the Second World War.

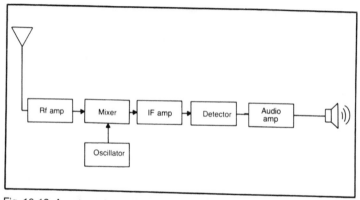

Fig. 10-18. Armstrong later came up with the superheterodyne circuit that is still the basis for radio receivers today.

(I-F). This frequency (very often 455 kHz) is the amplified. From there, it passes to a diode detector, then to the audio amplifier. See Fig. 10-19.

TRANSMITTERS

Now we turn our attention to transmitters, and go back to the old Armstrong regenerative receiver (Fig. 10-16). If too much energy were fed back to the input, the circuit would begin to oscillate on its own. When this happened, the signal produced was much steadier than anything produced by even the finest spark transmitter. However, in order to be received, the detector receiving it had to be oscillating also. It was tuned slightly off the transmitted frequency so that the difference between the received signal and the oscillation of the detector was within the audible range. This produced the pure, musical note that we now identify with radio telegraph dots and dashes.

The modern superheterodyne receiver has a built-in oscillator that produces this note when it receives code signals, and provides a carrier for single-sideband voice reception.

Sometime between the two world wars, quartz crystals came into use for frequency control. When an electric signal is imposed on a section of quartz crystal, it produces a mechanical strain. The instant that strain is released, the crystal vibrates for a few microseconds and that vibration causes it to deliver a minute alternating current at the frequency of the vibration. This frequency is dependent upon the mechanical dimensions of the crystal fragment. Therefore, if the crystal is placed anywhere in the feedback circuit

170

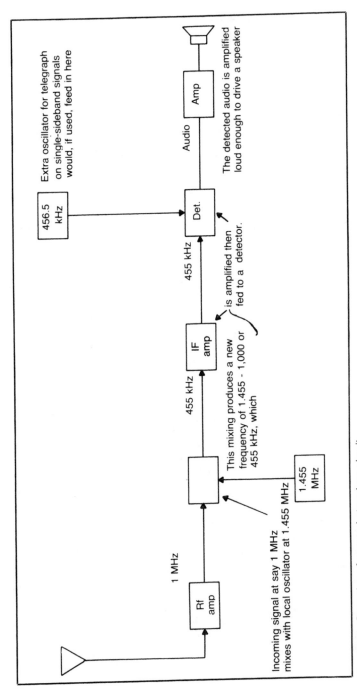

Fig. 10-19. Operation of a superheterodyne circuit.

Extra oscillator for telegraph on single-sideband signals would, if used, feed in here

The detected audio is amplified loud enough to drive a speaker

Audio

...is amplified then fed to a detector.

This mixing produces a new frequency of 1.455 - 1,000 or 455 kHz, which

Incoming signal at say 1 MHz mixes with local oscillator at 1.455 MHz

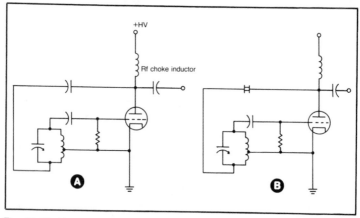

Fig. 10-20. How a crystal can be added to an oscillator circuit to control its frequency.

of an oscillator, it will control the oscillator frequency. Figure 10-20A shows an oscillator circuit and its crystal-controlled equivalents, Fig. 10-20B.

Until the introduction of transistors, the standard transmitter arrangement consisted of an oscillator followed by one or two amplifier stages. Some were crystal controlled; others used a variable-frequency oscillator (VFO).

Voice transmission was achieved by simply amplifying the voice to a power level equal to the transmitter output power, and connecting it through a transformer in series with the output amplifier, Fig. 10-21. It was an expensive way to do it but, for a long time, the only practical way.

Transmitters now are quite different. For one thing, they use transistors instead of vacuum tubes (with the possible exception of the output stage). Also, instead of amplitude modulation (except in broadcasting and CB), they produce a single-sideband signal.

The signal is generated at a relatively low frequency and mixed with the audio signal in what is called a balanced modulator. This produces two sidebands with no carrier. One of the sidebands is filtered out, leaving a low-frequency, single-sideband signal. This signal is then mixed with the signal from another oscillator to produce a higher-frequency signal in the same manner as two signals are mixed in a superheterodyne receiver. When the output signal is at the proper frequency, it is amplified and applied to an antenna. Figure 10-22 is a block diagram of a single-sideband transmitter.

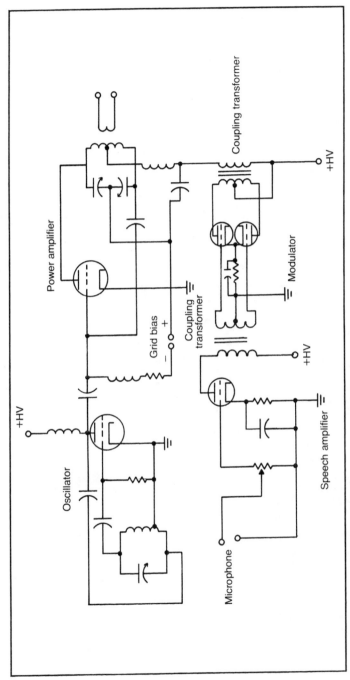

Fig. 10-21. The combination of circuits that make an AM radio transmitter.

173

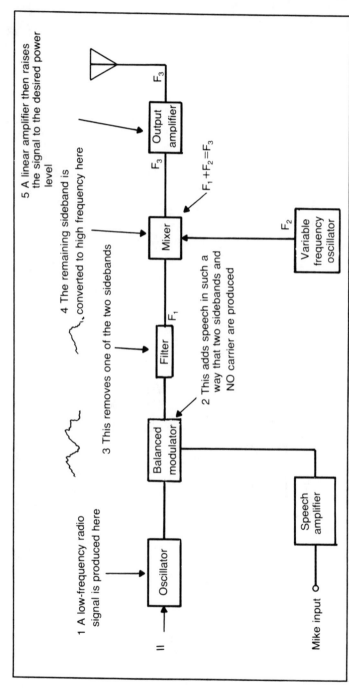

Fig. 10-22. Block diagram of a single-sideband radio transmitter.

It would be possible to write an entire book on radio transmitters and receivers, but what we have covered is more than enough for one chapter. For further study, there are a number of excellent TAB books on the subject.

TELEVISION

Television uses all the techniques of radio communication, along with photographic techniques and a couple of other instruments. There are two transmitting systems used. One is a frequency-modulation (FM) transmitter no different from those used in FM broadcasting. This carries only the sound channel. The other produces a single-sideband AM signal with the carrier present. This carries all the picture information. See Fig. 10-23. The picture signal contains so much information to be transmitted so quickly (30 complete pictures per second) that it occupies six times as much of the radio spectrum as the entire AM broadcast band.

The picture signal is very complex. It contains the actual information of the picture elements, vertical and horizontal synchronizing information, and color information. To better understand all this, let's look at it a bit at a time.

First the actual picture information. The TV camera focuses the picture on a photo-sensitive screen which develops an electric charge that is proportional to the amount of light, and precisely distributed as an electrostatic image. This screen is inside a special kind of vacuum tube. The electron beam from the cathode of the camera tube is caused to sweep back and forth across the picture. This produces a varying dc output that is a precise electrical record of the light and dark parts of the picture as they were scanned by the beam in the camera tube.

The video signal from the camera modulates an AM transmitter. At the receiver, the signal is detected, amplified, and fed to the grid of the picture tube. The picture tube produces a cathode ray similar to that in the camera tube, which sweeps back and forth across a fluorescent screen. This causes the screen to light up, and the brightness is proportional to the intensity of the beam. Since the beam is modulated by the video signal, the picture tube produces light and dark areas that are exact duplicates of the light and dark areas in the camera tube.

It all works fine, as long as the tubes in the camera and the TV receiver scan the picture area exactly in step with each other. This is accomplished by means of synchronizing pulses imposed on the

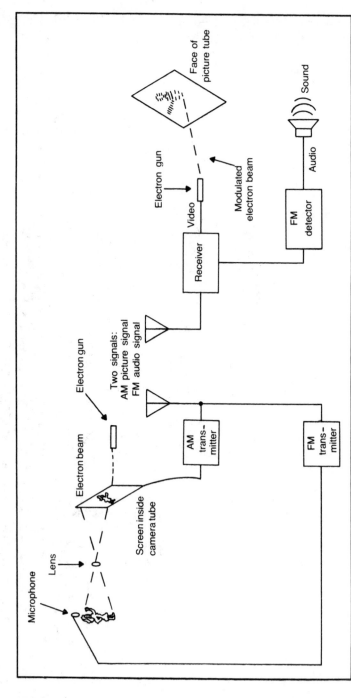

Fig. 10-23. The principles of a television system.

video signal. A signal containing both picture information and sync pulses is called a "composite video signal."

The television receiver, then, is a combination of an AM receiver, FM receiver, and a device called a cathode-ray oscillograph (the circuits that operate and control the picture tube). If color is involved, the complexity of the system triples. There are three sets of video information involved, together with a special color-information signal. Perhaps, when you realize all this, you may feel a bit sympathetic to your local TV repairman.

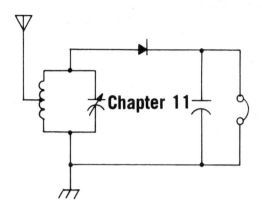

Digital Logic

There are two areas of electronic application: *analog* and *digital*. Analog circuits operate from smooth variations in voltage or current levels. Digital circuits know only two states—on or off.

Electronic calculators, computers, games, and a host of other devices use digital circuits. Their successful operation depends upon the proper switching and combining of these two states through a variety of numerous circuits until the desired result is achieved.

An early electronic computer was made by IBM, and used neither tubes nor transistors. It operated by the combining of thousands of electromagnetic relays. Shortly thereafter, tubes were used to do the switching electronically. Computers became slightly smaller. That is, they needed a smaller room to house them (about 30 by 50 feet).

Transistors reduced the size again to the point where a computer could occupy a single wall. Then the size went down to that of an office desk. Now, with micro technology, a computer with all the capabilities of the original IBM, and then some, can occupy no more space than a good-quality TV set.

Regardless of size, these devices, along with a host of other industrial applications, operate strictly from combinations of on and off functions. Let us now examine a few of the electronic devices that wind up being crammed into a computer.

TRANSISTOR SWITCHING

A transistor can be used to switch a circuit or device on or off just as if it were a relay. When a voltage is applied between the base and the emitter (negative with a PNP transistor, positive with an NPN transistor), the resistance between the collector and the emitter becomes very low. We therefore have an on/off function which is controlled by the base current. See Fig. 11-1.

Two transistors can be connected together so as to control one another. Such a circuit has two stable states in which one transistor is turned on and the other turned off. This is called a "flip-flop" circuit, see Fig. 11-2. As you can see from the circuit, the collector of each transistor is coupled to the base of the other. If transistor A is conducting, its collector voltage is close to ground. This keeps the base of transistor B at a low voltage. Transistor B is not conducting, so its collector is at a high voltage. Since it is coupled to the base transistor A, it keeps that base high, and ensures that it continues to conduct.

If the base of transistor B is momentarily forced high, B starts to conduct. This forces the base of transistor A down, cutting it off. The condition is now reversed—transistor B conducts and transistor A does not.

A flip flop is a special kind of circuit that was originally developed back in the vacuum-tube days. In those days it was called a multivibrator, and was of relatively little importance until the invention of computers. There are three kinds of multivibrators. They are called astable, bistable, or monostable.

An astable multivibrator is also called a free-running multivibrator, see Fig. 11-3. It is, in fact, a kind of oscillator. However, instead of producing a sine wave as do most other oscillators, it produces what is called a square wave. That is, it produces the kind of wave that might be made by simple, on/off switching. Such a signal is exactly the kind used in most computers and in all digital electronic devices.

A monostable multivibrator (Fig. 11-4) must be triggered by an external signal in order to operate. Then it switches momentarily and returns to its original state. In the digital electronic field, it is now commonly called a one-shot. A single trigger pulse will cause it to produce a single output pulse, the duration of which is determined by the circuit component values.

Bistable multivibrators (Fig. 11-5) are the most important of the three. These are the true flip flops, and there is a whole family of

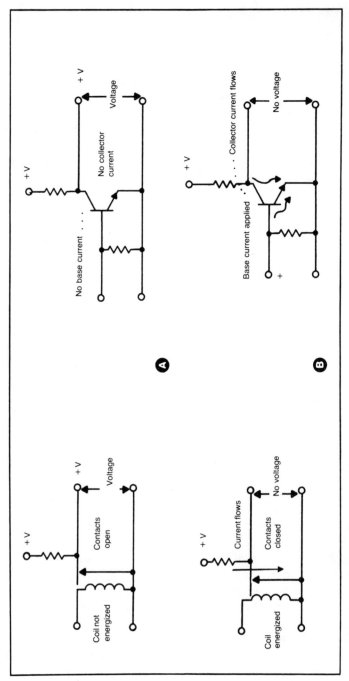

Fig. 11-1. A transistor switch and its relay equivalent.

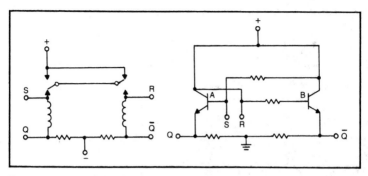

Fig. 11-2. A bistable flip-flop circuit and its relay equivalent.

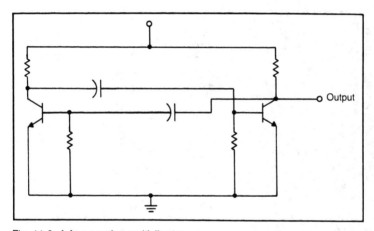

Fig. 11-3. A free-running multivibrator.

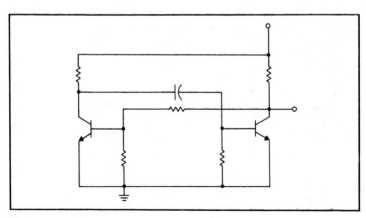

Fig. 11-4. A monostable, or one-shot circuit. An input trigger results in a single pulse out.

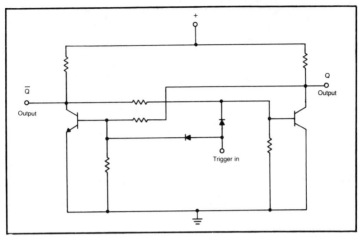

Fig. 11-5. A triggerable flip-flop, also called a divide-by-two counter.

them, each suited to a particular purpose. Some solid-state versions can be adapted to any of several possible applications merely by rearranging the external connections.

The basic bistable multivibrator is called a J-K flip flop. The term "flip flop" is generally used for this whole family of circuits, and is descriptive of its behavior. The two transistors are wired in such a way that when one is turned on, it holds the other off. The other transistor, by being turned off, holds the other on. When the proper input signal is applied, the two transistors switch. Number one turns off, forcing number two to turn on. They remain in that state until a signal is received to again switch them, see Fig. 11-6. It is a very simple device, yet forms the basis for the operation of an incredible variety of devices. Some of the variations will be discussed shortly.

GATES

Even simpler than the flip flop, but equally as important, are two devices in the general family of *gates*. There are two kinds of gates, AND and OR.

An AND gate has two or more inputs, and one output (Fig. 11-7). If both inputs have a voltage applied, the output will turn on. Otherwise there is no output.

The other kind of gate is the OR gate, Fig. 11-8. Like the AND gate, it has two (or more) inputs. A signal applied to either gate will produce an output, without affecting the other input.

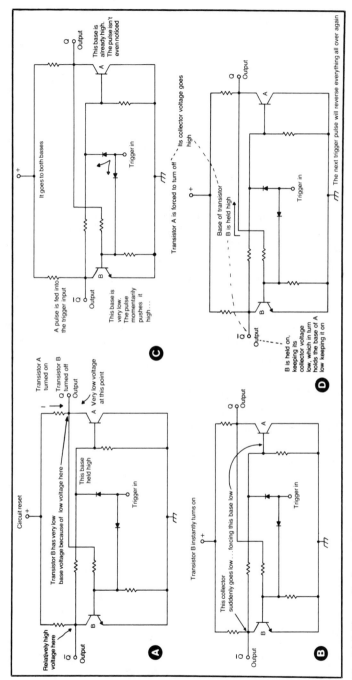

Fig. 11-6. How a flip-flop works.

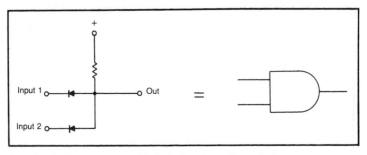

Fig. 11-7. AND gate circuit with its logic schematic symbol.

There are numerous other gates, such as the NAND gate, which is the opposite of the AND gate. It gives an output when all the inputs are *off*, but if one or more inputs come on, the output becomes zero. The NOR gate has zero output if all the inputs are on, and gives an output when any one or more inputs go to zero. Figure 11-9 shows NAND and NOR gates.

Finally, there is the inverter. This is a simple transistor switch. When the input goes high, the output goes low (Fig. 11-10).

All these devices seem very simple, and, in fact, they are. The complexity of modern digital electronics comes from the intricate combinations necessary to perform the various tasks.

AN EVENT MONITOR

Let us give you an example by describing the operation of a timing device (greatly simplified). The object is to determine if either of two events happens within a particular block of time. Let's say that we want to determine if it occurs no more than fifty

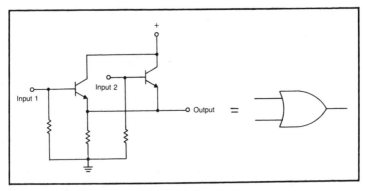

Fig. 11-8. OR gate with its logic schematic symbol.

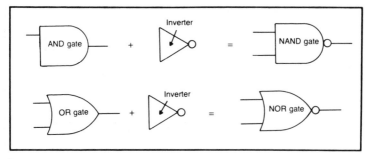

Fig. 11-9. By inverting their outputs, the AND and the OR gates are converted to a NAND and a NOR gate.

milliseconds after the start, and no less than thirty milliseconds. The digital circuits (collectively called *logic circuits*) are shown in Fig. 11-11.

First of all, you will notice that it is a block diagram: the parts are, except for a few, not shown. This is because most of the common logic devices come ready made as integrated-circuit "chips." Except for occasional peripheral components, integrated circuits are the only parts used.

Notice that there is a block called the clock. This is merely a crystal-controlled oscillator that gives a square-wave output. That is, the output is a rapid train of on/off pulses. To simplify the explanation, I have chosen a frequency of 1000 pulses per second.

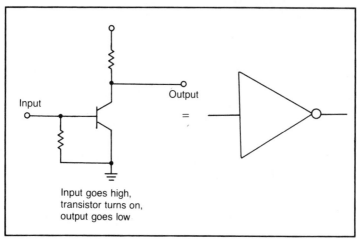

Fig. 11-10. An inverter is a simple transistor circuit that turns the signal upside down.

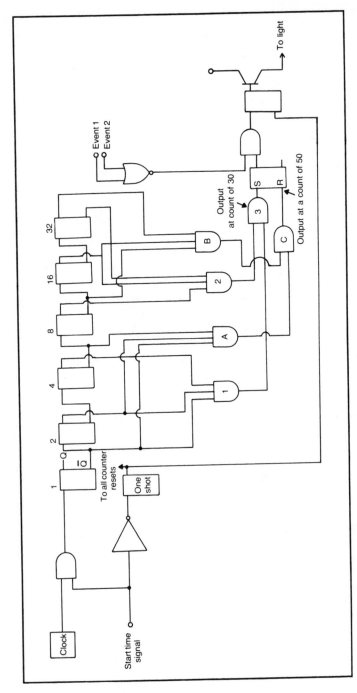

Fig. 11-11. The first time you see it, a logic circuit is rather frightening.

The output of the clock is connected to one of the inputs of an AND gate. The other input receives a positive voltage as a command to start timing. (Generally, either a positive or negative voltage can be called a logic *high*. The polarity depends on the kind of devices you are using. Zero volts is then called a logic *low*.)

When a logic high is applied to the second input of the AND gate, the output will go high and low in step with the clock output. The gate, therefore, effectively switches the signal from the clock, see Fig. 11-12.

The clock signal is "gated," as they say, with the start command and then fed to a chain of flip flops, Fig. 11-13. These flip flops are so wired that each successive pulse will change their state. Such a flip flop is sometimes called a *divide-by-two counter*.

Note that each flip flop has two outputs. One is called the Q output, the other is the $\overline{Q}$ (not Q). When the flip flop sets, the Q goes high, and the $\overline{Q}$ goes low. When it resets, the condition reverses. The $\overline{Q}$ of the first counter is connected to the input of the second. When the first pulse out of the clock reaches the first counter, the Q output goes high, and the $\overline{Q}$ goes low. The second clock pulse sends the $\overline{Q}$ high, and the positive transition trips the second flip flop. Thus it take two clock pulses to trip the second flip flop once. Figure 11-14 through 11-17 show this sequence.

If you now follow the succession of the circuit and give it a little thought, you will see that it will take four clock pulses to trip the third flip flop once. Then it will take eight clock pulses to trip the fourth, sixteen to trip the fifth, and thirty two or trip the fifth.

We are trying to measure milliseconds. Each clock pulse represents one millisecond. The first time interval we are looking for is thirty milliseconds. If you can follow the action of the devices, the 30th clock pulse will produce this condition: The Q output of the flip flops representing 2, 4, 8, and 16 will be high, the Q output of the rest will be low. (Note that these four numbers add up to thirty.)

The $\overline{Q}$ output on the flip flops that are reset is high. If we take all those points that are high on the count of thirty, and connect them to an AND gate, that gate will output only when the counter reaches the count of thirty, and at no other time. I suppose you could build a six-input AND gate, but it's easier to use the readily-available combinations of three and two inputs. At any rate, at the count of thirty, the gate spits out a positive signal. This pulse turns on, or sets a flip flop, see Fig. 11-18.

In a similar manner, we gate the points that will be positive at the count of fifty. In this case, it is the Q output of the flip flops

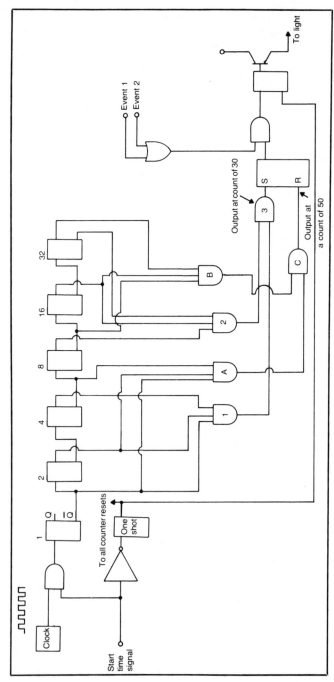

Fig. 11-12. Let's follow the events in the logic chain step by step. The clock signal cannot get through the AND gate because the "start" input has low voltage on it.

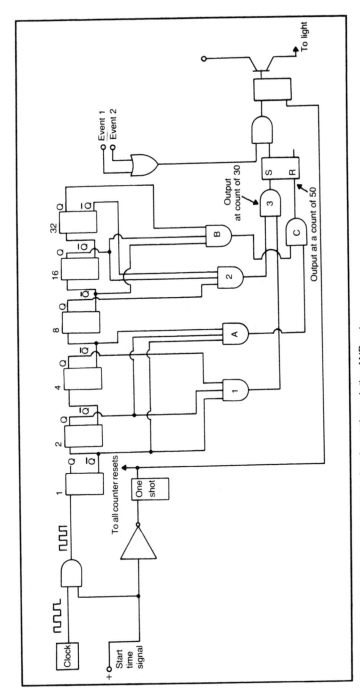

Fig. 11-13. When "start" goes high, clock signals get through the AND gate.

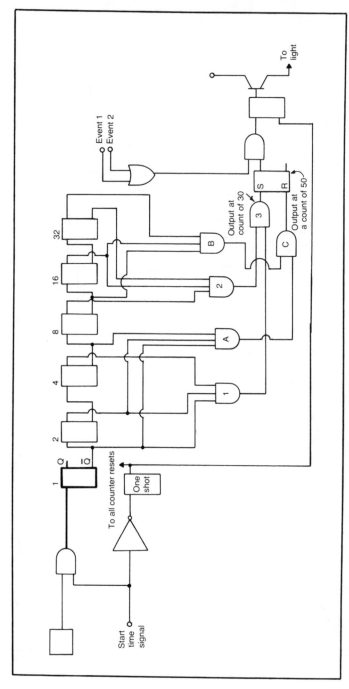

Fig. 11-14. At the first pulse, the Q output of the first counter goes high.

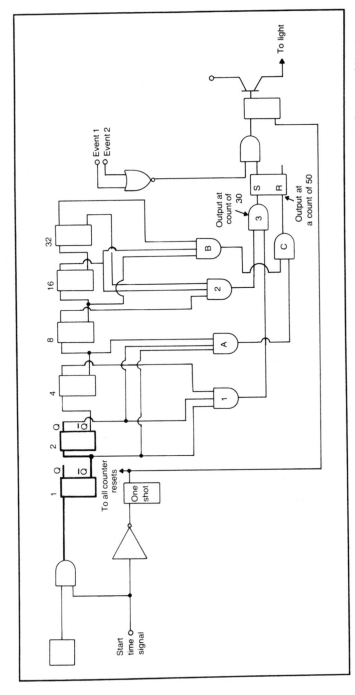

Fig. 11-15. At the second pulse from the AND gate, the Q output of the second counter goes high, triggered by the Q output of No. 1.

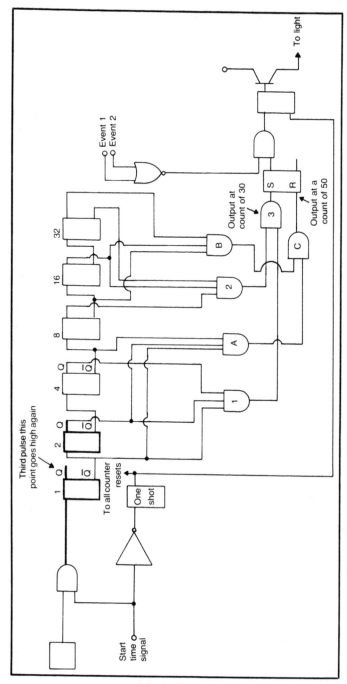

Fig. 11-16. At the third pulse, the Q output of the first counter is high again.

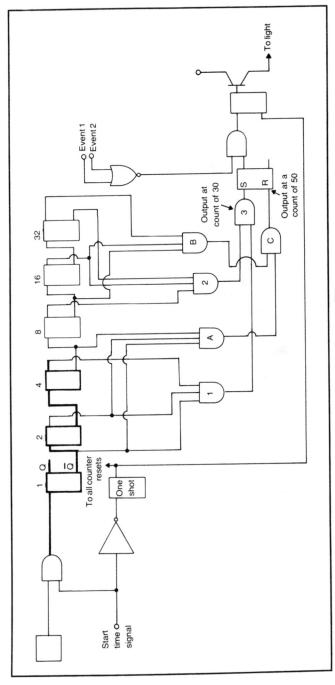

Fig. 11-17. On the fourth pulse, the first $\overline{Q}$ goes high, to trigger flip-flop 2, and the $\overline{Q}$ of that flip-flop triggers number 4, which causes its Q output to be high.

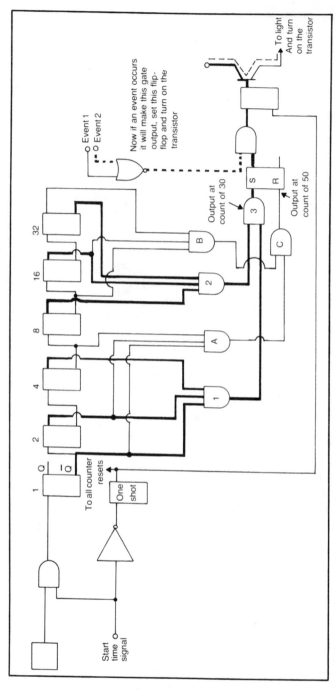

Fig. 11-18. After the 30th pulse, the Q outputs of flip-flop 2, 4, 8, and 16 are all high. This drives AND gates 1 and 2, which drives AND gate 3. This drives the S input of the last flip-flop. Its Q output then drives the AND gate to a condition ready to turn on if an event occurs.

representing 32, 16, and 2, and the Q outputs of the others. This pulse resets the flip flop, Fig. 11-19.

The flip flop now produces a positive pulse which exists only between the time of thirty and fifty milliseconds. This pulse goes into a two-input AND gate. The gate's other input is activated by the particular event you are trying to monitor. If the event occurs during the thirty to fifty millisecond "window," the gate will produce an output, otherwise, the gate will produce no output. The diagram shows two event inputs, feeding through an OR gate. My purpose in showing the two inputs that way is to show how an OR gate might be used if you are looking for either of two inputs. If you only have one input, the OR gate is not needed.

The output of the last AND gate operates a flip flop which turns on a transistor to drive a lamp. If the event occurs at no less than 30 or more than 50 milliseconds after the start signal, the lamp will light. Otherwise, it will not.

When the signal is removed from the "start time" input, the output of an inverter goes positive. This throws a "one-shot" multivibrator, the output pulse of which resets the counters and the lamp-driver flip flop, Fig. 11-20.

It seems quite complex, and it is. Each individual circuit is very simple, but the combination can be puzzling. If this operation was a thing that was to be built thousands of times, it might be practical to produce it as a single integrated-circuit chip. However, every chip that is on the market was first developed by individual building blocks such as the hypothetical one we have just analyzed. If you were to look at the logic in one of the video game chips, it would be almost horrifying. It does show, however, that some pretty amazing things can be accomplished out of these basic building blocks by exercising a little patience and ingenuity.

BINARY NUMBERS

One purpose in showing you this particular example is to point out the value of a clock and counter. This combination can be found in a wide range of electronic devices ranging from digital watches to computers. The handy feature of the flip-flop circuit is a natural for all kinds of mathematical operations, but the two-state operation of this device is best applied using the *binary* system of numbers.

The binary system is a way of expressing numerical quantities using only two numbers, one and zero. It takes a greater number of digits to express any given quantity than with our decimal system, but that drawback is more than made up by its ease of use in

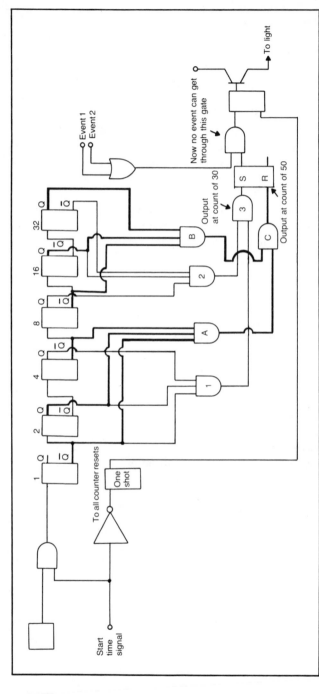

Fig. 11-19. After the 50th pulse, the Q outputs of flip-flops 2, 16, and 32 are high, and the Q of 1, 4, and 8 are also high. These outputs are used to drive AND gates A and B, which drive C. This resets the last flip-flop, which turns off the final AND gate, thus blocking any event from acting on the light circuit.

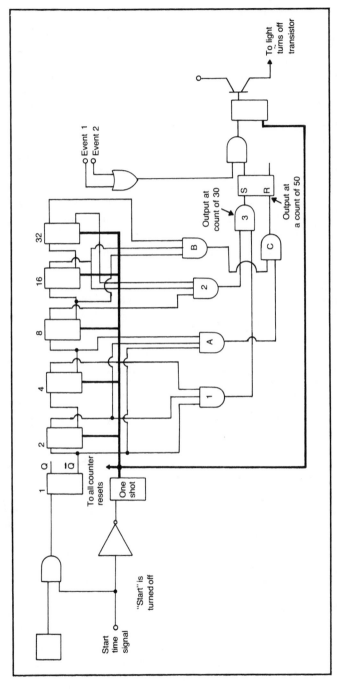

Fig. 11-20. The entire counter is reset by turning off the start circuit, which causes the inverter to present a high to the one-shot. This delivers a pulse to all counter reset lines.

electronics. Binary is, in fact, the basic language of all digital electronic devices.

The system of numbers we customarily use is based on the number 10. Each digit represents a power of ten (1, 10, 100, 1000, and so on.) In the binary system, each digit represents a power of two. That is, "10" is the number 2, "100" is the number 4, "1000" is the number 8, and so on. Each place to the left of the decimal point represents a double of quantity. There are only two numbers, 1 and 0, as shown in Table 11-1.

Let's look it over. The first digit represents a quantity of either nothing or one. If we already have one and add one more, we must use the second digit, and call it "10." This means one quantity of two and no more. One more than that is expressed "11" (one quantity of two and one more, which, of course, is three). To express four, we go to a third digit, and write "100." Five is "101," six is "110," seven is "111," and then eight demands a fourth digit, "1000."

Once you understand the progression of digits in the binary system, you can add, subtract, or do any mathematical function with them. Now let's see why they adapt so easily to electronic devices.

Table 11-1. Binary Numbers.

0	000000	Each time a number is doubled, you just put a zero on the end.
1	1	For example
2	10	3 = 11 6 = 110
3	11	10 = 1010 20 = 10100
4	100	15 = 1111 30 = 11110
5	101	Odd numbers end with 1
6	110	Even numbers end with 0
7	111	
8	1000	
9	1001	
10	1010	
11	1011	
12	1100	
13	1101	
14	1110	
15	1111	
16	10000	
17	10001	
18	10010	
19	10011	
20	10100	
21	10101	
22	10110	
23	10111	
24	11000	
25	11001	
26	11010	
27	11011	
28	11100	
29	11101	
30	11110	
31	11111	
32	100000	

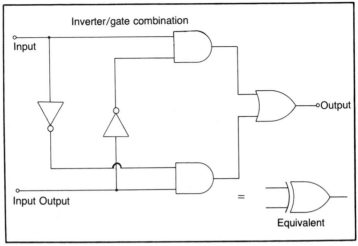

Fig. 11-21. Two AND gates and one OR gate make an exclusive-OR gate.

The flip flop, you will remember, has two states, set and reset. In the set state, the Q output has full voltage present, and in the reset state the Q output is zero. (The $\overline{Q}$ output is exactly the opposite.)

Let's use the Q output in the example. We can call the *set* condition (high) binary one, and the *reset* condition (low) binary zero. At the start of the count in the demonstration circuit, all flip flops are reset. That is, all the Q outputs were zero. On the first count, the Q output of the first flip flop goes to high (1). On the next count, it goes low, and the Q output of the second flip flop goes high. Thus, if the first flip flop is on the extreme right, and the others to the left of it, the state of the six of them can be expressed as 000010, which is the binary representation for two. On the next count, they become 000011, and on the one after that, 000100. The fifth count sets the first flip flop again, and the reading becomes 000101. Next, it will become 00110, then 000111, and 001000. Expressing the high of the Q output as 1 and the low as 0 the circuit was literally talking in binary.

Remember that we were seeking a count of 30. The number 30, expressed in binary digits, is 011110. We took the Q outputs of the 16, 8, 4, and 2 counters and gated them with the $\overline{Q}$ outputs of the 32 and 1 flip flops. Only at the count of 30 would all these outputs be high. Similarly, for the count of 50, we looked for highs on 32, 16, and 2, and we gated those with the $\overline{Q}$ (binary 0 of 8, 4, and 1. This combination is only present at the count of 50. We could, by select-

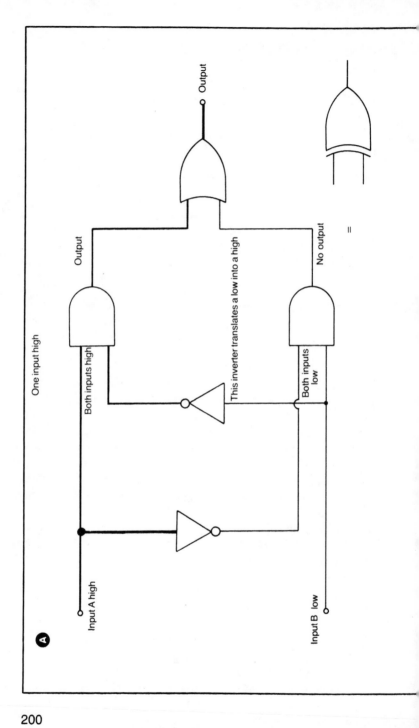

One input high

Input A high

Both inputs high

Output

This inverter translates a low into a high

Both inputs low

No output

Input B low

Output

=

A

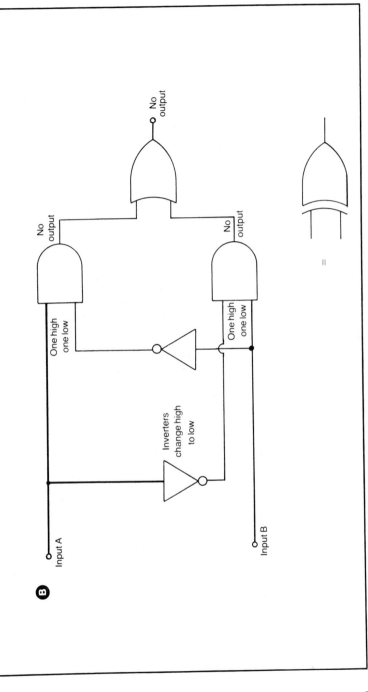

Fig. 11-22. Here's how an exclusive-OR gate works.

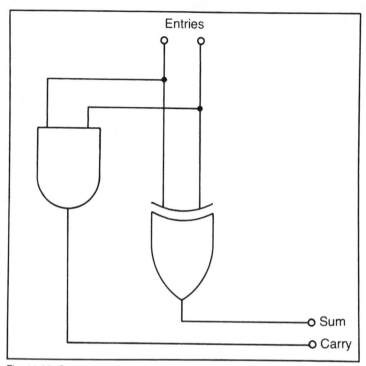

Fig. 11-23. Combine the exclusive OR gate with the AND gate, and you have a half adder.

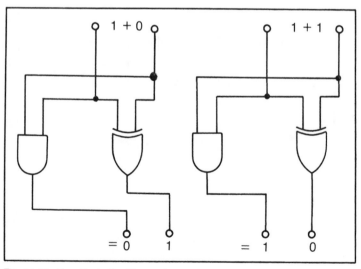

Fig. 11-24. How the half adder works.

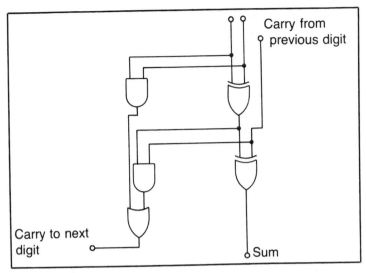

Fig. 11-25. Full adder circuit.

ing the combination of Q and $\overline{Q}$, set the circuit up for any numerical quantity from zero to sixty three. We would need a seventh flip flop to count 64 or higher.

EXCLUSIVE-OR GATE

One of the most important special devices in the digital-logic field is called the *exclusive-OR* gate. This is a device with two or more inputs and one output. Similar to the OR gate described previously, a voltage applied to either input can cause an output, but the difference is that only one input can be energized at a time. If more than one input is energized, there will be no output.

There are several ways to produce an exclusive OR function. Figure 11-21 shows an exclusive-OR gate.

Note that each input to the OR gate must first pass through an AND gate. The second input to each AND gate comes from the other circuit input through an inverter. Let's look at it in binary language. Remember that binary 1 means the input has voltage applied, binary 0 means the input is at zero volts. Let's see what will happen if input A is at binary 1 and input B is at binary 0 (Fig. 11-22A).

The voltage from input A must pass through the AND gate in order to get to the OR gate. It can only do that if both inputs to the AND gate are at binary 1. Since input B is binary 0, the inverter between input B and the input A AND gate turns the voltage into a binary 1 and applies it to the AND gate. Thus, a binary 1 can feed through to the OR gate and out into the subsequent circuits.

Now suppose both inputs are at binary 1 (Fig. 11-22B). Input A presents a high to the AND gate, but input B presents a low to the same gate so no output is possible. Input B presents a high to AND gate B, but the high on input A is inverted to a low to prevent output from AND gate B. Therefore, nothing gets through the circuit.

Since the circuit is symmetrical, if input A were low and input B were high, the B AND gate could deliver an output to the OR gate.

By using the exclusive-OR function, we can make a machine that can add. If we take an exclusive-OR gate, and put an AND gate with it, we have what is called a half adder (Fig. 11-23). This has two inputs, called A and B. Figure 11-24 shows the operation. If both are energized, the output of the exclusive-OR gate is binary 0. However, the AND gate outputs a binary 1. We have one digit that is 1 and one that is zero. This is binary 10, which is the number 2. The circuit has added binary 1 plus binary 1 to produce binary 10. If two half adders are cascaded as shown in Fig. 11-25, we have a full adder circuit. A four-digit adder would consist of a half adder for the units digit, and a full adder for the 2, 4, 8, 16, etc.

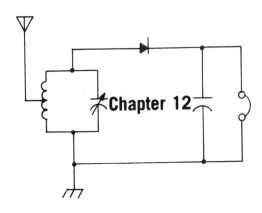

Computers

The computer is the most complex of electronic devices, yet the building-blocks of which it is made are the simplest. They are, in fact, the various gates and flip-flop circuits described in Chapter 11, connected into an incredibly huge and complicated array. The building-blocks themselves form modules, which in turn form a computer system. These modules are physically in the form of integrated-circuit chips, Fig. 12-1.

While the chip that forms the heart of the computer is physically about the length and width of a stick of chewing gum, the active electronic section inside is much smaller—the approximate size of a fingernail, as shown in Fig. 12-2.

Most computer systems can be boiled down to three sub systems: the Central Processing Unit (CPU), Memory, and interfaces to the Input/Output (I/O) devices. All of these sections are combinations of the digital devices described in the last chapter. However, modern manufacturers can build them incredibly small.

Here are some comparisons. The computer built in 1944 at Harvard University see Fig. 12-3. It consisted of over 3,300 electromagnetic relays. (Transistors weren't invented until 1948.) In 1948, the first truly electronic computer was built using vacuum tubes. Both these devices were housed in a large room, weighed several tons, and the vacuum-tube model required a full-time technician walking around behind it replacing tubes. The ENIAC, as it

Fig. 12-1. A microprocessor integrated circuit. (Courtesy Intel Corp.)

was called, had 18,000 of them, and was "down" more than it was operating.

It is only within the last 10 years that large-scale integration has been accomplished to the point where desktop-size computers have been practical. Consider this. A simple, four-function cal-

Fig. 12-2. Most of the IC is just to facilitate handling. The chip itself looks like this. (Courtesy Intel Corp.)

Fig. 12-3. The Mark I, a product of cooperation between IBM and Harvard's Dr. Howard Aiken, was the largest electromechanical calculator ever built. Completed in 1944, it had 3,300 relays and weighed 5 tons. The Mark I was basically a linkage of 78 adding machines and calculators. It operated for 15 years. (Photo courtesy IBM.)

culator has over 2,000 transistors in its chip. A microcomputer can easily contain over 30,000 devices. One advanced chip contains over 65,000, see Fig. 12-4. Considering the magnitude of technology in these devices it is almost saddening to see such devices used to move a colored spot around on a TV screen in one of the popular video games!

Computers are, however, used in much more productive applications, ranging from the control of a microwave oven or an automobile engine to the point-of-sale terminals functioning as what we used to call cash registers.

Without going too deeply into the details of the various modules, we can still acquire some understanding of the way these things work. As mentioned, the average system consists of the central processing unit, memory, and input/output devices. Let's look at these one at a time.

THE CPU

The central processing unit is the device that does all the hocus-pocus of the computer system. It is, in fact, the actual computer. It is here that the information is manipulated to get the desired result, it is here that the decision-making is done.

Fig. 12-4. Here is a microprocessor chip close up. The more orderly areas are memory registers; the jumbled-up areas contain the processing and decoding circuits. (Courtesy Intel Corp.)

The central processing unit contains, first of all an arithmetic-logic unit, which does the actual work. The ALU is supported by a half dozen or so temporary storage areas (registers) which hold data while the ALU is busy doing other things; by a command decoder, whose function is self explanatory; and by several other neat little circuits to keep up with the general housekeeping. Figure 12-5 is a simplified diagram of microprocessor chip architecture.

Computers talk in a language of binary numbers. Each digit (1 or 0 is a *bit*. The average home computer handles eight binary bits at a time. A group (or word) of eight bits is called a *byte*. Now, one byte (eight binary bits) can represent any number from zero to 256, depending on the particular combination of ones and zeros. To simplify the expressions, some systems divide each byte into two halves, and each half is expressed as a *hexadecimal* digit.

Hexadecimal (hex) means a base of sixteen. In hex numbering, each column can go, instead of the customary 0 to nine, from 0 to 15. A one in the second column and a zero in the first represents a quantity of 16. Table 12-1 shows how it works.

Once you understand hexadecimal numbers, you can mentally break down a byte into two four-bit groups by glancing at the hexadecimal value expressed. For instance, the expression "C3" is much easier to visualize in a complex program than 11000011. Yet they both represent the same digital encoding. For this reason, most machine-language programs are written in hexadecimal numbers.

The central processing units of most home computers have, in addition to numerous control connections, eight data connections, which can either input or output, and sixteen memory connections, which also can either input or output information. These two groups are known as the data bus and the address bus. We shall hear more about them shortly.

MEMORY

The memory-storage modules commonly consist of two kinds of memory. Some portion is called RAM, which means *Random-Access Memory*. Random-access memory is that kind into which information can either be stored or recovered. When you turn the power off, the information stored in RAM is lost.

ROM means *read only memory*. Data stored in ROM is there permanently, whether the power is on or off. The computer can read it out, but cannot store anything into it. It is here that the commands necessary to start up the computer operation must be stored. Some kinds of ROM, known as PROM, can be programmed by the user, if

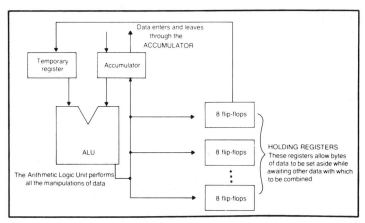

Fig. 12-5. The microprocessor has an arithmetic-logic unit, several registers, as well as program-counting and command-decoding circuits not shown here.

Table 12-1. Computers Are Programmed in Hexadecimal Numbers. Here's How They Work.

			Binary Expression 8 Bits, or 1 Computer Byte							
	Decimal number	Hexadecimal number	½ byte one hex digit				½ byte one hex digit			
	1	1	0	0	0	0	0	0	0	1
	2	2	0	0	0	0	0	0	1	0
	3	3	0	0	0	0	0	0	1	1
	4	4	0	0	0	0	0	1	0	0
	5	5	0	0	0	0	0	1	0	1
	6	6	0	0	0	0	0	1	1	0
	7	7	0	0	0	0	0	1	1	1
	8	8	0	0	0	0	1	0	0	0
	9	9	0	0	0	0	1	0	0	1
	10	A	0	0	0	0	1	0	1	0
	11	B	0	0	0	0	1	0	1	1
	12	C	0	0	0	0	1	1	0	0
	13	D	0	0	0	0	1	1	0	1
	14	E	0	0	0	0	1	1	1	0
	15	F	0	0	0	0	1	1	1	1
	16	10	0	0	0	1	0	0	0	0
	17	11	0	0	0	1	0	0	0	1
	18	12	0	0	0	1	0	0	1	0
	19	13	0	0	0	1	0	0	1	1
	20	14	0	0	0	1	0	1	0	0
2 × 16	32	20	0	0	1	0	0	0	0	0
3 × 16	48	30	0	0	1	1	0	0	0	0
4 × 16	64	40	0	1	0	0	0	0	0	0
5 × 16	80	50	0	1	0	1	0	0	0	0

he has the right equipment. Then the program is locked in until it is, in some types, erased, or it is stored forever in other types.

Whether it is RAM or ROM, each memory device can hold a certain number of bits. The more popular types have their contents organized into byte-wide groups. The number of bytes a chip can hold depends on its type and size. Figure 12-6 is a photomicrograph of a memory chip.

Whether it releases its information in groups of one, four, eight, or sixteen, there must be an organized way to select the group of bits the memory must release to the data bus. This selection is done by entering a numerical code (binary, of course) to select the particular group. This selection is known as the memory address. Since the memory address bus from the central processing unit consists of sixteen bits, it can address up to 65,536 addresses. A computer in which the chip is capable of reaching this number of locations is said to have a 64k memory capacity. Thus, when you

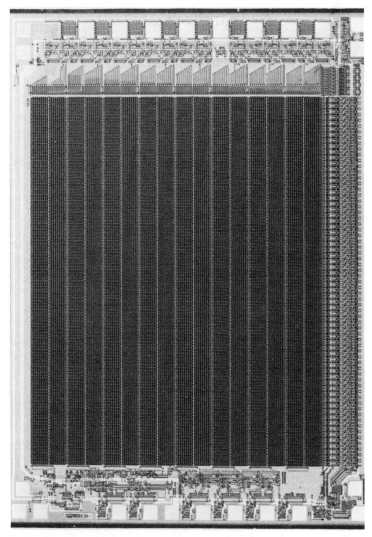

Fig. 12-6. Here's a close-up of a memory-storage chip. Note the orderly rows containing thousands of devices. (Courtesy Intel Corp.)

upgrade a home computer from, say 8k to 16k, all you do is add memory chips.

The microprocessor has, as we mentioned earlier, a memory bus of sixteen bits. This connects to the address inputs of the memory chips. When the microprocessor is turned on, an internal program counter causes the address bus to be scanned in binary numerical order, beginning with location zero and progressing until command to do otherwise. The contents of each location is read, and a command decoder within the microprocessor determines the meaning of the commands and causes the microprocessor to execute them. Exactly what the sequence of commands is determines the amount of intelligence the computer has. Now let's take a quick look into a microprocessor, Fig. 12-7.

THE ALU

The workhorse is the *Arithmetic-Logic Unit* (ALU). Here numbers are added, subtracted, or manipulated in other ways according to the commands.

Data usually comes in and out through a register called the accumulator. A register is a set of flip-flop circuits capable of retaining digital information. The average microprocessor has, in addition to the accumulator, half dozen or so additional registers where data can be temporarily set aside while the accumulator handles other traffic.

PROGRAMMING BASICS

For all its complexity, a computer can do nothing without program information. This is usually stored in memory, and circuits

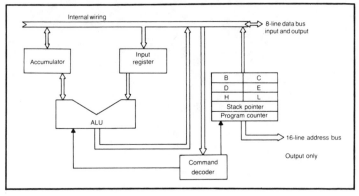

Fig. 12-7. A simplified diagram of microprocessor architecture.

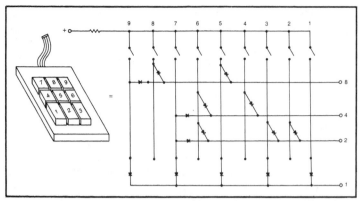

Fig. 12-8. Let's pretend that we have a keyboard wired up with this circuit. If you push a button, the output will be a binary equivalent of the key value.

within the microprocessor interrogate each memory location, one at a time, looking for instructions. The instructions must be exact, or the computer gets all fouled up. Moreover, since it is just a dumb machine, every last function must be spelled out. Otherwise, it'll either make a mistake or just sit there.

To illustrate just how a program works, let's imagine a simple system to accept two digits from a keyboard and add them up. While this system would work, it would, in the real world, be a hideous waste of good computer time. Still, it can serve as an illustration.

The system consists of a nine-button keyboard with a matrix of diodes to produce four-bit binary numbers, Fig. 12-8.

I have chosen Intel's 8080 or 8085 microprocessor, as these are the ones I'm most familiar with. Program is stored in a programmable ROM chip. For output, the data feeds through a latch into a BCD-to-7-segment decoder, which in turn drives an illuminated display. This hypothetical system is shown in Fig. 12-9.

There was a time when machine-language programming was a very tedious job. It still is, but not nearly as much so, thanks to other computers containing a program called an assembler. Assemblers take a series of special abbreviations (mnemonics), and translates them to the binary bytes that go into memory. They also assign specific memory locations and, in general, get the program ready to put into memory. Assembler programs can also print out a listing showing the memory address numbers, and the operation codes (both expressed as hexadecimal digits). The program to make the microprocessor in our imaginary system work is shown as an assembler printout in Fig. 12-10.

The first thing a programmer does is tell the assembler what memory addresses, or other information, will be accessed through input ports, output ports, or used for special temporary storage. In our listing, this is done in lines 5 and 6. Once this information is fed into the assembler, you only have to tell it to load from input or store to output, and the program will automatically call out the correct addresses when it assembles the routine. Then, as it is put into the new system's memory, all the correct addresses will be called out.

After listing the important access information, it is customary to begin a program by initializing the microprocessor (telling it to get ready). In our program, we will initialize simply by setting the accumulator and all the registers to zero. We could probably even get by without that. Lines 10 through 13 take care of initializing the system.

In the program listing, you notice a number of words followed by a colon, such as START:, or WAIT:. These are called labels, and inform the assembler program that, elsewhere in the new program, these particular addresses must be called up. They are only for the benefit of the assembler, to help it work. The only information that actually goes into the new system memory is the hex digits in the second column from the left. The extreme left column contains the memory addresses where these codes will be permanently stored. All the rest is commands given to the assembler, and comments for the programmer's personal use to help him keep track of where he is.

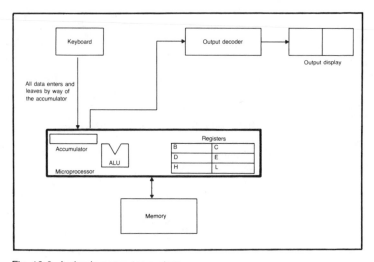

Fig. 12-9. A simple computer system.

```
ISIS-II 8080/8085 MACRO ASSEMBLER, V4.0        MODULE  PAGE   1


  LOC  OBJ      LINE      SOURCE STATEMENT

                    1 ;DEMONSTRATION TO ILLUSTRATE ASSEMBLY OF A SIMPLE PROGRAM
                    2 ;INPUT AND OUTPUT PORTS ARE ACCESSED THRU MEMORY ADDRESSES
                    3 ;
                    4 ;
  01FF              5 INPUT EQU 1FFH  ;INPUT PORT ADDRESS
  01F0              6 OUTPUT EQU 1F0H ;OUTPUT PORT ADDRESS
                    7 ;
                    8 ;INITIALIZATION
                    9 ;
  0000 AF         10        XRA    A    ;ZERO ACCUMULATOR
  0001 010000     11        LXI    B,0000H ;ZERO REGISTERS B AND C
  0004 110000     12        LXI    D,0000H ;ZERO REGISTERS D AND E
  0007 210000     13        LXI    H,0000H ;ZERO REGISTERS H AND L
                   14 ;
                   15 ;AFTER INITIALIZATION, WE BEGIN THE ACTUAL PROGRAM
                   16 ;
  000A 3AFF01     17 START: LDA    INPUT  ;LOAD A FROM INPUT (ADDRESS 01FF)
  000D CA0A00     18        JZ     START  ;JUMP IF ZERO TO START (000A)
  0010 47         19        MOV    B,A    ;MOVE A INTO REGISTER B
  0011 3AFF01     20 HOLD:  LDA    INPUT  ;LOAD A FROM ADDRESS 01FF
  0014 C21100     21        JNZ    HOLD   ;JUMP IF NOT ZERO TO HOLD (0011)
  0017 3AFF01     22 NEXT:  LDA    INPUT  ;LOAD A FROM INPUT (ADDRESS 01FF)
  001A CA1700     23        JZ     NEXT   ;JUMP IF ZERO TO NEXT (ADDRESS0017)
  001D 80         24        ADD    B      ;ADD ACCUMULATOR TO REGISTER B
  001E 27         25        DAA           ;CONVERT THE ANSWER TO BCD
  001F 32F001     26        STA    OUTPUT ;SEND THE ANSWER TO OUTPUT (ADDRESS 01F0)
  0022 3AFF01     27 WAIT:  LDA    INPUT  ;LOAD A FROM INPUT (ADDRESS 01FF)
  0025 C22200     28        JNZ    WAIT   ;JUMP IF ZERO TO WAIT (ADDRESS 0022)
  0028 C30A00     29        JMP    START  ;JUMP TO START (ADDRESS 000A)
                   30        END


PUBLIC SYMBOLS


EXTERNAL SYMBOLS


USER SYMBOLS
HOLD    A 0011   INPUT  A 01FF   NEXT   A 0017   OUTPUT A 01F0   START  A 000A WAIT
                                                                            A 0022
ASSEMBLY COMPLETE,  NO ERRORS
```

Fig. 12-10. This is an actual assembler readout of the program for our imaginary computer system.

Let's now examine the program one memory location at a time. Beginning at the first location, on line 10 in the program, we find the command AF. This is the hexadecimal code for 10101111. When this code is received, the computer sets the accumulator to zero.

The mnemonic XRA A tells the assembler program to load this command into the first available location. Note that a zero is written "Ø". The slash through it distinguishes it from a capital O.

On Line 11, we have the command LXI B, ØØØØH. The op code is Ø1 ØØ ØØ. Note that this is a three-byte command (in Hex). Therefore it occupies three memory addresses. It loads zeroes into register pair B and C. Lines 12 and 13 do the same to the remaining registers.

Now the microprocessor has all its registers set to zero. In a larger program, a module called the *stack pointer* would also have to be set to some RAM address, but since this program uses no RAM, we don't bother with the stack pointer.

A microprocessor works very fast. Each command takes, on the average, five machine operations per byte of command. However, the 8085 chip performs three million machine operations per second. It could whip through this whole program thousands of times in the brief time it takes you to push one button on the keyboard. Therefore, we must have some means of holding it back until a button is pushed.

Look at Line 17 in the program. Beginning at address ØØØA, (hexadecimal for address No. 10) we have a three-byte command, 3AFFØ1. This causes the microprocessor to load the accumulator with whatever's in address Ø1FF. (Note that, in a command containing an address, the last byte of the address is given first and the first byte last.) The ROM containing the program has no such address as Ø1FF. However, the address is wired so that this address will connect the data bus to the input port from the keyboard.

In the brief time that the data bus is connected to the keyboard, the keyboard will either be giving it a number or it will be a zero. If the keyboard is zero, the command in line 18 sends the program counter back to Memory Location ØØØA (labeled START). Thus the microprocessor will continually interrogate the keyboard port until you push a button. See Fig. 12-11. When the input is not zero, it will then and only then go on to the next command.

Let's suppose that we now have pushed a button, Fig. 12-12A. Just for the argument, let it be a number 5. In line 17, which the microprocessor has repeatedly returned to, the microprocessor loads this number into the accumulator. (Line 18 tells it to jump back *if* the accumulator contains a zero.) Now, however, it does not. It contains a 5 (ØØØØ Ø1Ø1). Therefore, it ignores Line 18 and goes on to line 19. Line 19 is memory address ØØ1Ø (hex for 16), and this contains the hex code 47 (Ø1ØØ 111Ø). The mnemonic MOV B,A

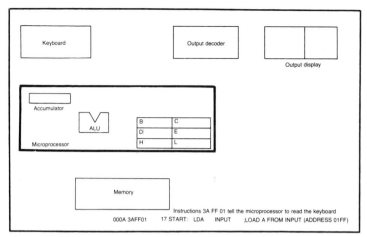

Fig. 12-11. The program causes the computer to examine the input repeatedly until a number is entered.

gives us a hint as to what this means. It moves, or more accurately, it duplicates the contents of the accumulator into register B. Now, if a new number is put into the A accumulator, we still have the first number preserved in the register, Fig. 12-12B.

Remember, however, the machine can go through many thousands of commands in one second. It has read the port and safely stored your number in the register, but you are still pushing the button. We have to hold things up long enough to get your finger off the first button and onto the second button.

To do this, we have the microprocessor keep looking at the input (line 20, labeled HOLD:). Line 21 tells the microprocessor to jump if the input is *not* zero, back to the memory address we have labeled HOLD. Thus it will keep looking at this port and doing nothing until you finish pushing the button, several thousand operations later. See Fig. 12-13A.

We come to the label NEXT at line 22. Here we repeat what was done at the start of the program. We keep looking at the input port, and jump back if there's a zero.

You again push a button. Let's just suppose it's a seven. Instantly, the accumulator contains Ø000 Ø111. Since it's not zero, the program counter jumps to the next command, line 24, address Ø01D. There we have the code 8Ø (1000 Ø000). The mnemonic is ADD B. It causes the microprocessor to add the contents of register B to the contents of the accumulator. In binary, Ø101 plus Ø111 equals 11ØØ. See Fig. 12-13B. This is binary for 12, but hex for "C."

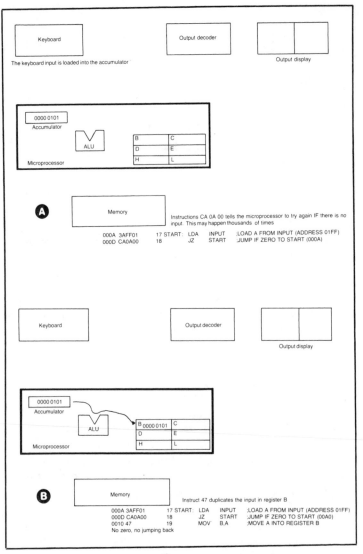

Fig. 12-12. Then the number is stored in one of the registers.

On line 25, the hex code 27 (ØØ1Ø Ø111) converts this to what is called *Binary-Coded Decimal*, Fig. 12-14. The contents of the accumulator change to ØØØ1 ØØ1Ø. Note that it is two digits, binary one and binary two.

Line 26 tells the microprocessor to store the contents of the accumulator into memory address Ø1FØ, which just happens to be the address that outputs whatever's on the data bus to the readout, in this case, an inexpensive pair of chips that convert the BCD code to the necessary combination to light up a seven-segment readout, as in Fig. 12-15.

All this happens while you're still pushing the button for the second digit—a tiny fraction of a second. Now, in line 27, we go back to the familiar combination of interrogating the input and jumping back if it's not zero. As soon as it's zero, the last line causes the program counter to jump back to the beginning.

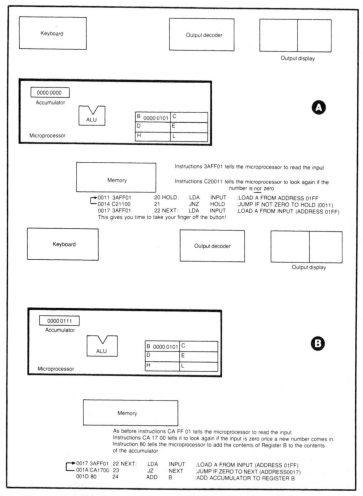

Fig. 12-13. Then it checks the input until the number you pushed goes away (A). Next it begins looking for the second number (B).

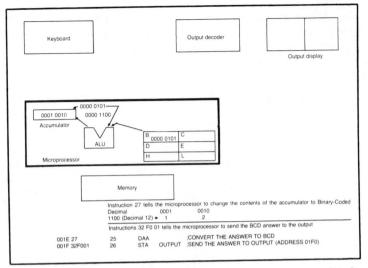

Fig. 12-14. When the second number comes in, the program causes it to be added to the number already stored, and converts the sum to Binary-Coded-Decimal digits.

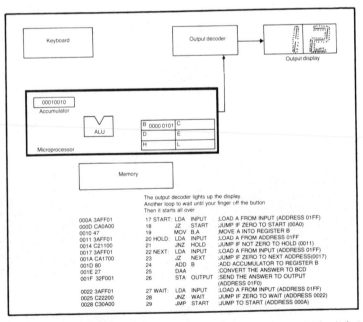

Fig. 12-15. The number is sent, in Binary-Coded-Decimal form to a circuit that decodes it and lights up the display. Then the program causes the computer to go back to the beginning and starts over.

It hardly seems worth it—all that just to add five and seven. That was why we said at the start that it would be a hideous waste to do it this way (unless you want to as an experiment). However, more complex programs are set up in a similar way to manipulate all kinds of information that can be encoded as binary numbers and, in turn, operate anything from a microwave oven to an "intelligent typewriter" or a word processor. See Figs. 12-16 and 12-17.

THE BASIC LANGUAGE

If you were to buy a home computer, you would not be expected to program it in this way. Home computers use a very high-level language called *BASIC*. This is translated into the machine language by a program. To add five to seven with BASIC, you would simply type "PRINT (5+7)" on the keyboard.

BASIC is one of the easier computer languages to learn. In fact, the average twelve-year-old can master it well enough to write very satisfactory programs. While each make of computer has its own version of BASIC, they are sufficiently similiar that, once you've mastered one, the others are very easily understood.

A BASIC program consists of a number of statements, consecutively numbered. The statements are usually numbered by tens

Fig. 12-16. Here is an office word-processing system. The program is stored in one of the disk drives at the operator's right, and the material she is typing in the other. The words appear on the TV screen, and she can change them around or correct spelling until she gets it the way she wants it. Finally, on command, the computer sends the document to the printer on the table behind the operator where it is printed out. The computer can be used for many other purposes simply by changing the program. (Courtesy Xerox Corp.)

Fig. 12-17. The Radio Shack TRS80 model III is a compact system that is inexpensive enough for small businesses and some home applications. Its capabilities far outstrip the monster shown in Fig. 12-3. (Courtesy Tandy Corp.)

(10, 20, 30, and so on), but they can be numbered in any increments you wish. The convenience of increments greater than one is that you can add statements in between existing statements whenever you find it necessary. A BASIC program may look something like this:

```
10   PRINT "HELLO. WHAT'S YOUR NAME?"
20   INPUT;N$
30   PRINT 'HELLO"N$ "HERE IS A MATH PROBLEM
     FOR YOU"
40   RANDOM: A=(RND 100): B=(RND 100)
50   PRINT "HOW MUCH IS "A" PLUS "B"?"
60   INPUT;C
70   IF C= (A+B) THEN 110
80   PRINT "SORRY "N$ "TRY AGAIN: INPUT D
90   IF D=(A+B) THEN 110
100  PRINT "WRONG AGAIN," N$ "THE ANSWER IS" (A
     B):GOTO 40
110  PRINT "VERY GOOD" N$ "NOW TRY ANOTHER
     ONE":GOTO 40
```

This is a very simple program, designed to assist a third grader

222

in arithmetic. It should work, with a little modification to satisfy the individual BASIC language of computer used. Once the program is loaded into memory, the operator starts it simply by typing the word "RUN."

Immediately, the computer translates the wording of line 10 into the necessary machine language to make it flash the words "HELLO. WHAT'S YOUR NAME?" onto the screen. Then line 20 requires it to wait for an input, which is labeled N$. Throughout the program, the string of letters labeled N$ will be substituted for the symbol. That is how the computer appears to recognize the operator by name.

Line 40 causes the computer to select two numbers between 0 and 100 at random, and labels these numbers A and B.

Line 50 puts the problem onto the screen, using actual numbers in place of the letters A and B. Then the operator is required to input an answer.

Line 70 makes a decision. If the operator enters a correct answer, the computer program jumps to line 100. It commends the operator for giving a right answer and selects a new problem by jumping the program back to line 40.

If the answer given is incorrect, it informs the operator (in line 80) that his answer was incorrect, and gives him another chance, still addressing him by name.

If a second incorrect answer is given, line 100 tells the correct answer, and jumps the program back to line 40 to begin a new problem. The actual operation is easy enough to learn that a third grader will get plenty of practice in math before the novelty wears off.

BASIC is a good, general-purpose language for the tinkerer, the small-business man, or for anybody who wants to whip up a quick, simple program. However, industrial users have other, more powerful languages for their computers. Two of the more popular ones are FORTRAN and COBOL. (All these names are acronyms alluding to the application intended.) FORTRAN (Formula translation) is the programming language used in scientific research calculations, and COBOL (Common Business-Oriented Language) is used in business bookkeeping.

These high-level languages can only work if the computer is programmed to translate them down to the machine language. Such translations are known as *compilers*.

Ready-made programs, called *software*, are generally recorded on a disk or cassette tape in machine language, although some may be recorded in the higher-level language with a translator included. Such software is generally intended for use in the computers that interface with a human operator. Computers that perform automated, robot functions, and only interface with machines, have their software stored in ROM.

And so we conclude our book with the general state of the art as of 1981-82. However, with the rapid progress of industry, no innovation, even in the very near future should be a surprise. Before this book becomes obsolete, we may well expect computers to be as common in every household as pocket calculators—perhaps more so.

Appendix
Where to Get the Goodies

For static-electricity equipment, this store deserves special mention, as it is a one-of-a-kind store to the best of my knowledge. In addition to assorted static machines, magnets, and electrical-experiment kits, they supply equipment to the other sciences that makes them an experimenter's delight. Their catalog is a superb amateur scientist's wishbook.

Edmund Scientific
101 Gloucester Pike
Barrington, NJ 08007

For electronic experiment supplies in general, the Radio Shack stores in most towns of any size have a pretty good stock, including learner kits for home study. If you don't know where the nearest one is, write to

Radio Shack
1600 One Tandy Center
Fort Worth, TX 76102

For make-it-yourself kits, and for an extensive line of accredited home-study courses try

Heath Company
Benton Harbor, MI 49022

AMATEUR RADIO

To learn how you can get into Amateur Radio, contact

American Radio Relay League
Newington, Connecticut 06111

There are a number of magazines dedicated to this hobby, and each also supplies books and study material. One of them is named *QST*, and it published by the American Radio Relay League at the above address.

The others are:

73	*Ham Radio*
Peterborough, NH 03458	*Greenville, NH 03048*

CQ
Cowan Publishing Co.
Port Washington, NY 11050

COMPUTER MATERIALS

Companies selling computer parts and materials are numerous, and I'm not too sure who's the biggest. A good swap sheet listing many sources of new and used materials is

Computer Shopper
P.O. Box F
Titusville, FL 32780

FOR FURTHER STUDY

We recommend the following excellent TAB Books:

TAB No.	Title
628	*Basic Electricity & Beginning Electronics*
1233	*AC/DC Electricity & Electronics Made Easy*
588	*Basic Electronics Course*
891	*Practical Solid State DC Power Supplies*
510	*How to Read Electronic Circuit Diagrams*
1113	*Understanding Electronics*
1186	*How to Install Your Own Stereo System—2nd Edition*
1137	*The First Book of Electronic Projects*
1083	*Practical Handbook of Stage Lighting and Sound*
973	*How to Be a Ham—2nd Edition*
1136	*Practical Electronics Math*
1182	*The Complete Handbook of Radio Receivers*
1338	*Basic Electronics Theory—with projects & experiments*

Glossary

acceptor—Impurity added to semiconductor material which causes it to become P type.

accumulator—The portion of a computer that receives data for processing and produces the final result of the processing.

acoustic—Referring to sound waves.

adder—An electronic system capable of adding binary numbers.

address—In computer work, a location in memory, or the particular combination of digits that accesses that location.

admittance—The ability to conduct alternating current.

alignment—The process of tuning many different circuits to make them work together, as in a radio or television receiver.

alternating current—Electrical current that periodically reverses. (The abbreviation ac also refers to a voltage of similar characteristic).

alternator—A mechanical device that produces alternating current.

amateur—One who practices an art just for fun. The word is no implication of a person's level of skill. Also capitalized to indicate a radio Amateur operator or the Amateur Radio Service.

ammeter—A device that measures current.

ampere (abbreviated amp)—The amount of current that flows when one volt is applied across one ohm. Also defined as one coulomb per second.

amplify—To raise the power level of a signal.

amplitude—The loudness of a signal.

amplitude modulation (abbreviated AM)—Varying the loudness of one signal to cause it to carry another.

analog—That branch of electronics in which operation is a function of the level of a signal.

AND gate—An electronic device, having two or more inputs, that delivers an output only when all inputs are active.

anode—The positive electrode in a vacuum tube or diode.

antenna—An electrical conductor, usually raised overhead, that collects radio signals.

arc—A luminous electrical discharge through space or through a gas.

armature—The moveable element in an electromechanical device.

arithmetic-logic unit (abbreviated ALU)—That part of a computer that does the addition or subtraction, or otherwise combines bits of information.

arrester—A device that protects electric communication wires from lightning.

ASCII—American Standard Code For Information Interchange. A digital code used in computer and communications work.

assembler—A computer program that organizes the coded instructions for storage into memory.

assembly language—The most basic language of computer instructions. All other languages used to program computers must be translated down to assembly language in order to work.

astable—That type of digital circuit which constantly changes state.

attenuate—The opposite of amplify. To reduce the level of a signal.

audio—That range of frequencies which, if converted to sound, can be heard.

audion—The name of an early amplifying vacuum tube.

autotransformer—A transformer that has only one winding.

ballast—An inductor connected in series with a fluorescent lamp.

band pass—An expression that refers to the range of frequencies accepted by a device.

base—The controlling element of a junction transistor.

BASIC—A popular computer programming language.

solid-state semiconductors, the negative element of a diode.

baud—The rate at which a digital device delivers a series of data bits.

beta—The current-transfer ratio of a transistor.

bias—A controlling voltage or current applied to a transistor or vacuum tube. Also an alternating current applied to a tape-recording head.

binary—A number system having a base of 2.

binary-coded decimal—The representation of decimal digits as binary numbers.

binaural—An early name for two-channel stereo.

bistable—Having two stable states.

bit—A single piece of digital data, usually represented as an on (1) or off (0) state.

black light—Another name for ultra-violet.

block diagram—A diagram of an electronic device in which the different modules are represented as boxes, rather than showing the connections of the individual parts.

boot—Short for bootstrap; a portion of a computer program that initializes the whole system and sets it up for the rest of the program.

breadboard—A technique of assembling a circuit so that all parts are readily accessible for testing and debugging. So called because the earliest ones were made by fastening the parts to a wood base resembling a cutting board.

bridge—1. An electrical circuit with several branches so wired that the behavior of current in one branch is an indication of adjustments made in the other. 2. A rectifier circuit consisting of four diodes so arranged as to use both halves of the ac cycle.

bug—A semi-automatic telegraph key.

BX—Flexible, metal-sheathed electric power cable.

byte—A series of data bits processed by the computer as a single piece of information. Usually 8 bits.

capacitance—The power-storing capability of a capacitor.

capacitor—A device consisting of two conductors separated by insulating material.

carbon—A black, non-metallic element that conducts electricity.

carrier—That portion of a radio signal whose only function is to help in demodulation at the receiver end.

cathode—That portion of a vacuum tube that emits electrons. In

cell—A single unit in an electric battery, capable of operating independently of the others.

character—In computer work, a specific, recognizable combination of bits that has a unique meaning.

circuit—A closed path from a power source, through a load, and back to the source.

circuit breaker—An electromechanical device that automatically interrupts a circuit in the event of overload.

clock—In digital electronics, an oscillator that produces a very stable, square-wave signal.

coax—Short for coaxial cable. Two-conductor cable in which both conductors have a common axis. That is, one conductor is wrapped around the other.

COBOL—A computer-programming language used for business and bookkeeping operations.

code—Any combination of characters or electrical functions that has a specific meaning.

coherer—An early form of radio detector in which the internal material coheres, or sticks together, when excited by a radio signal.

coil—A winding of electrical conductor.

collector—That electrode in a junction transistor through which the primary flow of current leaves the device.

common—That part of a complex circuit through which all branches ultimately return.

commutator—That portion of a motor or generator through which power reaches the armature windings.

comparator—An electrical device that compares two signals.

compiler—A computer program that generates the machine-language command codes from instructions entered by the programmer.

component—In an electronic device, a specific part such as a resistor or capacitor. In a system, any one of the devices that help make up that system.

condenser—An old name for a capacitor.

conductance—The ability of a material to pass electrical current. Opposite to resistance. The unit of conductance is the mho.

Conductor—Any material that permits the flow of electrical current.

connector—Any of a number of devices for electrically attaching two conductors.

continuous wave—Abbreviated CW. Radio jargon for communication by telegraph code.

coulomb—The quantity of electricity that passes a given point in a conductor in one second while a current of one ampere is flowing.

CPU—Abbreviation for Central Processing Unit; The heart of a computer system.

CRT—Abbreviation for cathode-ray tube, which is another name for a TV picture tube. Also used to describe a computer terminal using a video screen.

crystal—Any of a number of solid materials in which the atoms are arranged with some degree of geometric regularity. A piece of quartz crystal used for its peculiar electrical characteristics, as in an oscillator circuit.

current—The flow of electrons through a medium.

cycle—The change of alternating current through two complete reversals, back to the starting state.

CW—Short for continuous wave.

Darlington amplifier—A transistor amplifier circuit in which the emitter of one transistor directly feeds the base of another.

dash—The longer of the two basic elements used in radio-telegraph code.

data—Coded electrical information, especially as used in computers.

dB—Abbreviation for decibel, a unit of power or voltage ratio.

dc—Abbreviation for direct current.

debug—The process of detecting and correcting errors in an electrical circuit or in a computer program.

dead—Electrically inoperative.

decade—A factor of ten.

decimal—Based on a quantity of ten.

decrement—To count downward.

detector—A device that detects a signal.

dielectric—The insulating medium between the two electrodes of a capacitor.

digit—A single numerical character.

Digital electronics—That branch of electronics that operates solely from on/off transitions.

diode—A two-element vacuum tube, or the solid-state equivalent. A device that allows current flow only in one direction.

DIP—Abbreviation for Dual Inline Package. A form of integrated-circuit packaging in which the connections are arranged in two parallel rows, generally with the pins of each row 1/10 inch apart.

dipole—A center-fed antenna with two elements of equal length positioned end-to-end.

direct current—Electrical current that never changes direction.

disk—A disc-shaped medium for storing a computer program.

distortion—The departure of an output signal from the desired or from the characteristics of the input signal.

dot—The shorter of the two elements used in telegraph code.

drain—The electrode of a field-effect transistor that corresponds to the collector of a junction transistor.

drift—The gradual departure of a signal from the desired characteristic.

dry cell—An electric cell in which the liquid contents are sealed from access.

dynamic ram—A form of read-write computer memory.

E—The mathematical symbol for electrical voltage.

electromotive force—Another word for voltage, or electric pressure.

electrode—Any of a number of devices that provide necessary access to electric power.

electrolyte—A chemical substance that is acted upon by electricity.

electrolytic capacitor—A capacitor in which one plate and the dielectric are formed by an electrochemical action.

electrolysis—The stimulation of a chemical reaction by electric current.

electromagnet—A device in which magnetism is temporarily created by electrical action.

electromagnetic spectrum—The overall range of all possible frequencies of alternating current or electrical radiation, ranging from radio waves through radiant heat, visible light, X-rays, gamma rays, and beyond.

electron—One of the elementary constituents of matter, characterized by a negative charge.

electronic—That branch of electrical science that deals with the electrical action of vacuum tubes, transistors, and similar devices.

electroscope—A device that measures electric potential by means of the mechanical force between two charged bodies.

electrostatic—Pertaining to stationary electric charges.

element—Any of 104 substances that cannot be broken down to a combination of two or more substances.

emission—Any waves radiated into space by a device.

emitter—That electrode of a junction transistor which corresponds to the cathode of a vacuum tube.

enable—To render capable of operating. Turn on.

erg—The absolute gram/centimeter/second unit of energy and work.

exciter—That portion of an electronic system that supplies the earliest level of signal to be subsequently processed by the rest of the system.

fade—To diminish in signal strength, usually as a result of natural cause.

Fahnestock clip—A spring terminal intended for easy connecting and disconnecting of circuit elements.

farad—The unit of measurement of capacitance. If a one-farad capacitor is charged with one coulomb, a potential difference of one volt will exist between its terminals.

FCC—Federal Communications Commission. That government agency having jurisdiction over radio communications and radiating devices.

female—That type of connector into which the mating unit fits.

ferrite—A compressed, powdered material, having known magnetic characteristics, used in the tuning elements of radio equipment.

FET—Field-effect transistor. A semiconductor device that operates from the effect of an internal electric field rather than by current through a junction.

filament—A very thin wire used as the incandescent element in a light bulb, and as the heating element in a vacuum tube.

file—In computer work, a program or set of related records.

filter—A network of electronic components that has calculated responses to selected frequencies.

flashover—A disruptive electric discharge, either through space or along the surface of an insulator.

flash tube—The light-producing element of an electronic photo flash system. Also used in a strobe light.

flip flop—An electronic circuit having two stable states, easily switched from one to another by the proper signal. It is used in many applications, including computer memory.

flutter—Rapid variation in amplitude of a signal.

flux—1. A material used in soldering to enable the free flow of liquid metal. 2. The intensity of a magnetic field.

FM—Abbreviation for frequency modulation.

FORTRAN—FORmula TRANslation. A high-level computer language used principally in scientific calculation.

free electron—An electron that is not bonded to a particular atom.

free-running—Referring to an oscillator, operating without any external controlling signal.

frequency—The rate at which a repetitive electrical function occurs.

fringe—The outer limit of reception distance.

front end—That portion of a radio or TV receiver through which the signal enters.

full-wave rectifier—A rectifier circuit that uses both halves of the alternating-current wave.

fundamental—The lowest frequency present in a complex waveform.

fuse—A safety device consisting of a strip of conductor with a low melting point. In the event of current overload, the fuse melts, opening the circuit.

gain—The amount of power increase an amplifying device gives to a signal.

galena—A bluish-gray, crystalline form of lead sulfide, once used as a radio detector.

galvanic—An old word for current resulting from chemical action.

galvanometer—An instrument for detecting and measuring low-level electric currents.

gang—To mechanically couple two or more controls together.

garbage—Slang for unwanted, meaningless signals.

gate—An electronic device in which two or more inputs affect one output.

gauss—The metric unit of magnetic induction.

generator—1. A mechanical device that produces electrical power. 2. An electronic device that creates a signal, usually for test purposes.

Germanium—A bluish-white metal widely used in semiconductor devices.

giga—A prefix meaning one billion.

googol—The quantity represented by a one followed by 100 zeros.

gram—The basic metric unit of weight, equal to the weight of one cubic centimeter of water.

graphics—The ability of a computer terminal to produce pictorial information on the video screen.

grommet—A rubber or plastic bushing used to insulate a hole in a metal panel.

ground—The voltage-reference point in an electrical apparatus. Not necessarily earth connected.

half adder—An electronic circuit that accepts two binary inputs and delivers a sum output plus a carry output.

half-wave rectifier—A rectifier circuit that uses only one half of the ac cycle.

ham—Slang for amateur. Originated from amateur performers who made their makeup from hamfat.

hardware—In computer terminology, the actual equipment.

harmonic—An integral multiple of a signal.

head—The device that either receives sound or video from its medium or places it into the medium.

heater—In a vacuum tube, the filament.

heatsink—A mounting base designed to carry away and dissipate heat.

helix—A spiral shape, often used in UHF antennas.

henry—The basic unit of inductance. One henry produces an induced voltage of one volt with a current variation of one ampere per second.

hertz—The unit of alternating current or radio frequency, equal to one cycle per second.

heterodyne—The process of combining two signals of different frequencies, resulting in a new signal.

hexadecimal—A number system based on 16, used in computer programming.

high-pass filter—A device that allows the passage of high frequencies while blocking lower frequencies.

hole—An electron deficiency in the valence structure of a semiconductor. A hole acts like a positively-charged body.

hologram—A three-dimensional picture produced by the action of laser light.

hookup—A slang word to describe the connections in a circuit.

hybrid—An electronic circuit using both vacuum tubes and transistors, or using both integrated circuits and transistors.

I—The mathematical symbol for electric current.

IC—Abbreviation for integrated circuit.

image—In radio communication, a signal frequency, other than the desired, that can produce the proper intermediate frequency.

impedance—The opposition to the flow of alternating current resulting from the resistance of the circuit combined geometrically with the inductive or capacitive reactance.

inductance—The property of a circuit that opposes any change in the existing current.

infra-red—Radiation wavelengths longer than that of red visible light, but shorter than radio waves.

input—The signal fed into a device, or the terminals of a device intended to receive the signal.

insulator—Any of a number of substances through which electric current cannot flow.

integer—Any whole number.

integrated circuit—A number of transistors, diodes, and related devices built in a single piece of semiconductor material.

interface—The means of passing a signal between two or more devices.

interlock—A safety device that removes power from electrical equipment whenever the protective cover is removed.

intermediate frequency—Abbreviated i-f. A frequency to which a signal is converted in either the transmitter or receiver system.

interpreter—A computer program that converts a high-level language to machine code.

interruptor—A mechanical device that rapidly interrupts current.

inverter—An electronic device that produces an output exactly opposite to the input.

jack—A female circuit connector.

JFET—Abbreviation for junction field-effect transistor.

juice—Slang for electric power.

jumper—A short length of wire used for a temporary connection.

junction—The connection between two dissimilar types of semiconductor.

key—A hand-operated encoding device, such as a telegraph key.

kHz—Abbreviation for kilohertz.

kilo—Prefix meaning thousand.

L—Mathematical symbol for inductance.

label—A code name used in a computer assembler program.

landline—Slang for telephone.

laser—A device that produces an extremely coherent and intense beam of light (Light Amplification by Stimulated Emission of Radiation).

latch—A two-state circuit that holds in the activated state until it receives a deactivation signal.

lag—The condition whereby one factor of a signal is following another by a measurable amount of time.

lead—Opposite to lag.

leakage—Undesired flow of current through an insulator.

Leyden jar—The forerunner of the capacitor, named for Leyden, Holland, where it was invented.

light—Radiation of wavelengths to which the eye is capable of reacting, or closely related thereto.

lightning arrestor—A device containing a spark gap that by-passes harmful electric currents resulting from a lightning discharge.

linear—Having an output that varies exactly as the input.

load—That part of a circuit that consumes the power.

logic—That branch of electronics that deals with digital switching functions.

loop—A complete electric circuit.

loss—The undesired diminishing of power.

LSB—Abbreviation for Least Significant Bit or for Lower Sideband.

Luminaire—A complete lighting unit.

machine code—An operation code that can be recognized by a machine.

machine language—The operation codes for a microprocessor or computer.

magneto—An ac generator, usually for producing high voltage for automotive ignition.

main—That portion of a power distribution system, extending from the source, from which the distribution branches extend.

make—To close a circuit (as opposed to break).

male—That type of circuit connector that fits into another.

mark—A closed-circuit condition, often called binary 1.

maser—Same as a laser, except that its radiation is in the microwave region.

matrix—A coding network.

mega—Prefix meaning one million.

memory—The equipment and media to store computer machine-language information in electrical or magnetic form.

micron—One millionth of a meter.

microphone—A device that receives sound waves from the air and converts them to electrical impulses.

micro—A prefix meaning one millionth.

microwave—A term applied to radio signals at frequencies of 1000 MHz and above.

mil—One onethousandth of an inch.

milli—Prefix meaning thousandths.

mismatch—A condition in which the impedance of a load is not the same as that of the source.

mixer—A device that combines two or more signals.

mnemonic—A computer-code abbreviation.

modem—(MOdulator-DEModulator); a device that conditions digital codes for transmission over conventional communication systems.

modulate—To impose information onto a signal by varying amplitude, frequency, or other parameter.

molecule—The smallest particle of a substance that still retains all the properties of the substance.

monitor—1. To listen continually to a communication channel. 2. That part of a computer program that inputs information, displays it, and performs general operating tasks common to all programs.

monochrome—In reference to a TV picture, black-and-white.

monostable—Having only one stable state.

Morse code—The telegraph code originally devised by Samuel F. B. Morse. Today it is referred to as American Morse. International Morse code is a variation of the original.

mother board—A large printed-circuit board that interfaces between others.

multiplex—Sending two or more signals simultaneously via one channel.

nano—Prefix meaning one billionth.

negative—The type of electrical charge characteristic of a surplus of electrons.

neon—An inert gas used in some illuminating devices.

net—Short for network. An organization of communications stations.

neutral—Neither negative nor positive.

neutralize—To provide feedback through an amplifying system in opposition to the gain, thereby preventing oscillation.

Nichrome—A nickel-chromium alloy used in resistance wire and heating elements.

NiCd—Short for nickel-cadmium. A type of sealed, rechargeable battery cell.

Nicad—Trademark for NiCd cells made by Gould, Inc.

nixie—A glow-tube used in alphanumeric displays prior to the development of the light-emitting diode.

noise—Unwanted dynamic disturbance in an electronic system.

NOT—An expression referring to the digital opposite. In written notation, it is expressed by a line above the expression. For example, $\overline{Q}$.

null—A condition of minimum signal.

octal—A vacuum-tube base having eight pins.

octave—A differential of a full musical scale, equal to a doubling of frequency.

ohm—The standard unit of resistance measurement. One volt sends a current of one ampere through one ohm.

one shot—A name for a monostable multivibrator.

op amp—Operational amplifier. An integrated-circuit amplifier the gain of which can be set by the addition of external components.

optical—Pertaining to the behavior of light.

optimum—The ideal condition for best results.

orthicon—The tube in a television camera that converts the picture into electrical impulses.

oscillate—A circuit used to produce alternating current without moving parts.

oscilloscope—A device that produces a graph of a signal on a screen similar to a video screen.

overload—A load too great for a device to handle.

parallel—The connection of two or more devices in such a way that current can flow through either independently of the other.

passband—The range of frequencies that will pass through a circuit with nearly equal gain or loss.

patch cord—A cord or cable with a connector on either end, used to connect two or more devices temporarily.

peak—The high point of the amplitude of a signal.

peak-to-peak—An alternating-current measurement of the extreme excursions of current or voltage.

PEP—(Peak Envelope Power); the average power of a radio signal applied to the antenna during the highest crest of the modulation envelope, average over one cycle.

pentode—A vacuum tube having five elements, not necessarily including the filament.

period—The time interval in one complete cycle of repeating events.

permeability—A measurement of how well a substance can contain magnetic force.

persistence—The amount of time a picture tube continues to glow after the electron beam has passed.

pF—Abbreviation for picofarad; one trillionth of a farad.

phase—The relationship between two signals or components of a signal.

photocell—A device that produces electric current when acted upon by light.

pi—A mathematical constant, roughly equal to 3.14159.

piezoelectric—The property of producing electric voltage when mechanical stress is applied.

pigtail—A splice made by twisting the ends of two or more wires together.

plate—The anode of a vacuum tube.

polarization—The plane of oscillation of a wave.

positive—Opposite of negative. Electric force characteristic of a deficiency of electrons.

potentiometer—A variable resistance, characteristic of a volume control.

potential difference—Voltage.

preamplifier—An amplifier that raises a signal to a level suitable for processing without degrading the signal quality.

primary—The input winding of a transformer.

primary cell—A non-rechargeable battery cell.

printed circuit—A base for electric or electronic equipment in which all interconnecting wires are in the form of conducting lines etched on a phenolic or fiberglass base.

program—A sequence of operating instructions, usually automatically executed.

propagation—The manner in which radio waves are distributed after leaving the antenna.

pulse—A brief, abrupt change in a signal.

push-pull—An amplifier configuration in which two tubes or transistors work alternately, each taking one half of the signal cycle.

Q—A measure of the efficiency of a tuned device, expressed in whole numbers.

Q signals—A code commonly used in commercial and Amateur Radio communication. Examples:

> QRM - interference
> QRT - shut down
> QSO - communication
> QSL - acknowledgement
> QTH - home location

quad—An antenna consisting of a number of square loops.

R—Symbol for resistance.

radar—Acronym for RAdio Detecting And Ranging.

radiate—To give off a signal or energy.

radioactive—Producing energy of extremely short, potentially dangerous wavelengths.

RAM—Random-Access Memory. Computer memory that can receive and deliver data.

reactance—The opposition to the flow of alternating current.

readout—The visible delivery of data from a system.

Read-Only Memory (ROM)— Computer memory in which the data is permanently stored, and cannot be rewritten.

rectifier—A device that converts alternating to direct current.

reflection—Bouncing off a surface.

refraction—Being bent by passing between two media.

register—A short-term storage circuit.

relay—An electromagnetic switch.

repeater—In Amateur Radio, a station that receives signals on one frequency and retransmits them at another.

repel—Push away.

reset—To restore to the starting condition.

resistance—The opposition to the flow of electric current.

resistor—An electronic component having a fixed, known amount of resistance.

resonant—Having the maximum response to a particular frequency.

rf—Radio frequency.

ripple—Periodic fluctuation.

rms—Abbreviation for root-mean-square.

S meter—A meter on the panel of a communications receiver that measures signal strength.

St. Elmo's Fire—A visible electric discharge sometimes seen on the masts of boats or on the wingtips of aircraft.

saturation—The state of a transistor after which further increase of base current fails to increase the collector current.

sawtooth wave—A signal that changes at a linear rate between two fixed values.

scanner—A radio receiver that periodically samples several channels, one at a time, until a signal is received.

Schmitt trigger—A bistable circuit in which output is present at full value when the input exceeds a predetermined value, and is absent under all other conditions.

SCR—Abbreviation for silicon controlled rectifier, a solid-state switching device.

secondary—The output winding of a transformer.

selectivity—The ability to reject unwanted signals close to the desired signal frequency.

Selenium—A chemical element with strong photosensitive properties.

semiconductor—Any of several elements whose electrical properties are neither those of a conductor nor an insulator.

series—Connected in such a manner that current must flow through each device in turn, and all current flows through each device.

servo—An automatic control system.

shield—A grounded metal covering.

shift register—A number of flip-flop circuits so connected that data moves from one to another with each successive clock pulse.

short circuit—A circuit of much too little resistance for the source to handle.

shunt—Parallel connection. Often applied to a low-value resistance connected in parallel with an ammeter to extend its scale.

sidebands—Signals above and below an AM carrier frequency, resulting from mixing the carrier frequency with the voice signal.

signal—Any kind of electrical current or impulse that has a specific meaning.

Silicon—An element widely used in semiconductor devices.

simplex—A communication system in which communications are carried on in one direction at a time.

sine—A mathematical function determined by dividing the side of a right triangle opposite the related angle by the hypotenuse.

sine wave—An alternating-current wave that varies in amplitude according to a sine function. A pure signal without any harmonic elements present follows a sine curve.

single sideband—A mode of radio communication in which all the transmitter power is used to transmit the intelligence, the carrier being supplied at the receiver.

skin effect—The tendency for high-frequency signals to travel only in the outer surface of a conductor.

skip—The phenomenon of atmospheric reflection of radio signals by an ionized layer.

solenoid—A hollow magnetic winding that, when energized, draws a plunger inside.

solid state—The whole general field of rectifying and amplifying devices excluding vacuum tubes.

sonar—A detection system that operates by reflected sound waves.

sounder—An electromagnetic device that makes a tapping sound when energized.

source—The device that supplies power or signal to a circuit.

spaghetti—Flexible insulation, designed to slip over a wire.

spectrum—The overall range of usable radio or alternating current frequencies.

square wave—A signal that changes abruptly from one extreme to the other.

squelch—A circuit that silences a radio receiver when no signal is coming in.

standard—Any of several devices whose properties are known very accurately, and used to calibrate measuring devices.

static—Fixed, or unmoving. Also noise that interferes with radio reception.

stereo—A prefix meaning three dimensional.

storage battery—A rechargeable battery.

strobe—1. A bright, rapidly flashing light. 2. To rapidly examine or activate a large number of devices or circuits one at a time.

suicide cord—A length of two-conductor wire with an ac plug at one end and clips at the other. Not recommended, but widely used nonetheless in diagnostic testing.

superheterodyne—A radio receiver in which all incoming signals are converted to one intermediate frequency prior to the main amplification. Most radio receivers are superheterodynes.

suppressor—A resistance in the ignition wire of an automobile engine, used to suppress radiation or radio signals by the ignition spark.

sweep—To make a continuous change from one electrical value to another. The travel of an electron beam across the face of a cathode-ray tube.

SWR—Abbreviation for Standing Wave Ratio, a measure of the operating efficiency of an antenna.

synchronous—Operating in exact step with another device.

system—A number of electronic circuits working together to achieve a single result.

table—A collection of data, arranged for easy recovery.

tantalum capacitor—An electrolytic capacitor using Tantalum-foil electrodes, or Tantalum-oxide electrolyte.

taper—Rate of continuous change.

telegraph—An elementary electric communication system.

telephone—A system of transmitting voice over electric wires.

television—A system of electrically transmitting pictures.

temperature coefficient—The amount of change in a device resulting from temperature change.

terminal—A point of connection.

Tesla coil—An air-core transformer used to develop a high-voltage discharge.

tetrode—A four-element vacuum tube.

thermistor—A device that changes resistance at a known rate as a function of temperature.

thermocouple—A device that produces an electric voltage when heated.

toroid—A doughnut-shaped core used for special inductances and transformers.

track—1. The path of reproducible information in a recording medium. 2. To follow the path of recorded information. 3. To follow the tuning or adjustment of another circuit in exact step.

transceiver—A combination transmitter/receiver.

transducer—A device that delivers an electrical voltage when affected by sound waves or mechanical stress.

transformer—An electromagnetic device that passes alternating current from one circuit to another.

transient—A sudden change in conditions.

transistor—A semiconductor device capable of amplifying, oscillating, or transferring power.

TTL—(Transistor-Transistor Logic); integrated-circuit devices using multielement transistors.

transmitter—A device that sends out a signal.

trap—A device that suppresses undesired signals.

triac—A solid-state switching device, capable of switching alternating current.

trigger—A signal that initiates a series of events.

trimmer—A small, variable capacitor or resistor used to make fine circuit adjustments.

Tungsten—A brittle metal used to make the filament for electric lamps.

twin lead—A common, two-conductor ribbon cable used for lead-in wires of TV antennas.

twisted pair—Ham-radio slang for the telephone.

UHF—Abbreviation for ultra-high frequency.

ultra sonic—Sound waves of a frequency too high to be heard by the human ear.

ultra violet—That range of the radiation spectrum whose wavelength is shorter than violet light, and therefore invisible to the human eye.

unbalanced—A circuit in which one side is grounded.

vacuum—A space that is theoretically devoid of matter.

vacuum tube—An amplifying device in which electric current is controlled as it passes through a vacuum.

valve—An early name for a vacuum tube.

variable—Changeable.

VFO—Abbreviation for variable frequency oscillator.

video—Another name for a television signal.

volt—The unit of measure of electromotive force. One volt pushes one ampere of current through one ohm of resistance.

volt-ampere—A unit of apparent power in an alternating current circuit.

VU meter—A meter that measures speech power.

wafer—1. A thin, flat section of a rotary switch. 2. The piece of base material from which integrated circuits are made.

watt—A unit of measurement of power. The power of 1 watt produced when one volt moves one ampere of current.

white—Equal amounts of all components of the visible spectrum.

white noise—Electrical noise having components at all frequencies.

wire recording—A form of magnetic recording, antedating tape, in which the medium is a fine, stainless steel wire.

wireless—An early term for radio.

wire wrap—A technique for making solderless connections by tightly wrapping wire around a terminal.

write—In computer work, to place data into any kind of memory.

Xenon—A rare gas used in flashtubes.

xerography—The process of forming an image electrostatically.

X rays—A portion of the radiation spectrum having wavelengths far shorter than violet light, and great penetrating power.

xtal—Abbreviation for crystal.

xmtr—Abbreviation for transmitter.

Yagi—A mutli-element, directional antenna.

yoke—A set of windings that deflects the beam in a TV picture tube.

Zener diode—A semiconductor device deliberately designed to partially break down as the applied voltage exceeds a predetermined amount, and recover afterward.

Index

A

Acceptors, 7
Adaptor, 3-wire, 97
Alternating current, 76, 82, 125
ALU, 212
Ammeter, 39
Ampere, 39
Ampere, Andre, 59
Amplifier, radio-frequency, 169
Amplifier techniques, 144
Analog circuits, 178
Anode, 128
Armstrong, Edwin, 169
Atom, inner structure of, 3
Atom, the, 2
Attraction, electrostatic, 9
Auto transformer, 89

B

Bacon, Roger, 1
Ballast, 109
Base, 142
BASIC language, 221
Battery, basic, 29
Battery, simple, 30
Battery cells, antique, 30
Battery cells in parallel, 32
Battery cells in series, 32
Battery charger, 134
Battery charger filtering, 137
Binary-coded decimal, 218
Binary numbers, 195, 198
Bit, 208
Boxes, electrical, 101
Bridge circuit, 134
Buzzer, simple, 70

C

Capacitor, homemade, 18, 19
Cathode, 128
Cell, alkaline, 37
Cell, flashlight, 36
Cell, miniature, 38
Cell, storage, 33
Cell, voltaic, 27
Channel, 144
Charge, negative, 7
Charge, positive, 7
Charges, electrical, 4
Circuit, basic doorbell, 46
Circuit, parallel tuned, 164
Circuit, simple, 41
Circuits, basic, 96
Coherer, 157
Collector, 142
Compass, magnetic, 56
Computers, 205
Computer system, simple, 214
Conductance, 44
Conductors, 2
Copper, 28
Coulomb, 39
CPU, 207
Crystal phono pickup, 121
Current, 39
Current, direct, 26

D

Digital circuits, 178
Digital logic, 178
Diode detector, 168
Diode experiments, 126
Direct current, 26, 82, 125

Doorbell circuit, 70
Donors, 7
Drain, 143
Dufay, Charles, 2

E

Electrical charges, 4
Electric "bugs", 13
Electricity, static, 1, 8
Electric motor, 75
Electric motor, toy, 73
Electric power, 40
Electric sparks, 88
Electrochemical potential, 28
Electrolytic rectifier, 127
Electromagnetic induction, 76
Electromagnetism, 59
Electron, the, 2
Electrophorus, the, 11
Electroscope, a simple, 9
Electroscope, gold-leaf, 13
Electrostatic attraction, 9
Emitter, 142
Event monitor, 184
Exclusive-OR gate, 203

F

Filtering circuits, 136
Fixtures, electrical, 101
Flashlight cell, 36
Fleming valve, 127
Flip-flops, 183
Fluorescent lamp, 107
Franklin, Benjamin, 2
Full adder circuit, 203
Fuses, 102

G

Galileo, 1
Galvani, Luigi, 26
Galvanometer, 61
Gate, 142
Gates, 182
Generator, simple, 80
Gilbert, William, 1
Gray, Stephen, 2
Grid, 146
Grid-leak detector, 167
Ground, safety, 99

H

Half-wave rectifier, 131
Henry, Joseph, 60
Hertz, Heinrich, 156
Hexadecimal, 208
Holes, definition of, 42
Home wiring, 93

I

Induced back voltage, 78
Induced magnetism, 52
Inductive reactance, 92
Insulators, 2
Integrated circuit chip, 206

L

Lamp, electric, 98
Lamp cord, 95
Lamps, 106
Leyden jar, 15, 33
Leyden jar, home-brew, 17
Lightning, 24
Loads, electrical, 102
Logic, digital, 178
Logic circuit, 186

M

Magnes, 50
Magnesia, 50
Magnes stone, 50
Magnetic attraction, 51
Magnetic compass, 56
Magnetic deviation, maps of, 54, 55
Magnetic field, 56
Magnetic lines of force, 56-58
Magnetic principles, 51
Magnetic repulsion, 51
Magnetism, 49
Magnetism, induced, 52, 53
Magnetism, residual, 65
Memory, 209
Memory-storage chip, 211
Microphone, carbon, 118
Microphone, crystal, 121
Microphone, dynamic, 120
Microprocessor, 206
Miletus, 50
Monostable multivibrator, 179
Morse, Samuel, 60
Morse's original code, 72
Morse telegraph sounder, 67
Motor, basic, 71
Motor, electric, 75
Motor, series, 81
Motor, shunt, 81
Motors, 79
Musical scale, 112
Musschenbroek, Pieter van, 15

O

Oersted, Hans Christian, 59
Ohm, 39
Ohm's law, 39

P

Parallel circuits, 42
Permeability, 62
Phonographs, 124
Pith ball, 9
Poles of a magnet, 56
Power cord, 94
Power cord construction, 95
Power distribution system, 99
Power plug, replacing, 104
Primary winding, 85
Programming basics, 212

R

Radio communication, 155
Radio detectors, 128
Radiotelegraph code, 72
Radio transmitter, AM, 173
Radio transmitter, single-sideband, 174
Radio waves, 155
RAM, 209
Reactance, 162
Receivers, 165
Rectifier, half-wave, 131
Rectifier, full-wave, 133
Rectifier circuits, 133
Regenerative receiver, 169
Relay, simple, 69, 71
Resistance, 39
Resistors, series and parallel, 45
Resonance, 160
Resonant period, 160
Right-hand rule, 61, 62
ROM, 209

S

Safety grounds, 93
Schematic diagram, 41
Sealing wax, 11
Secondary winding, 85
Series circuits, 42
Sine waves, 83
Solenoid, simple, 65
Solid-state diodes, 131
Sound, 110
Sound, speed of, 111
Source, 142
Spark, climbing, 91
Spark coils, 88
Static electric experiment, 5
Static electricity, 8
Static electricity, industrial uses of, 22
Static electricity in nature, 25
Storage cell, 33

Strain relief 105
Sulphuric acid, 28
Superheterodyne circuit, 171
Symbols, electronic, 131, 132

T

Telegraph, 60, 66
Telegraph system, 69
Telephone, sound-powered, 123
Telephone, story of the, 113
Telephone reproducer, 115
Television, 175
Thales of Miletus, 1, 50
Transformers, 84
Transformer windings, 86
Transistor, field-effect, 142
Transistor, junction, 141
Transistor, npn, 143
Transistor, pnp, 143
Transistor amplifier circuits, 148
Transistor audio amplifier, 152
Transistors, 138, 140
Transistor switching, 179
Transistor types, 141
Transmitters, 170

U

Underwriter's knot, 105

V

Vacuum diode, 128
Vacuum tube, inside view of, 139
Vacuum tube, operation of, 140
Vacuum-tube amplifier experiment, 150
Vacuum-tube rectifier, 130
Vacuum tubes, 138
Voltage, 31, 39
Voltage drop, 43
Voltaic cell, 27
Voltmeter, 39

W

Watt, 39
Wire, joining pieces of, 77
Wiring, home, 93
Wiring types, 100

X

Xerox electrostatic copying process, 20, 21

Z

Zener diode, 149
Zinc, 28